Environmental Management Revision Guide

T0221346

The *Environmental Management Revision Guide: For the NEBOSH Certificate in Environmental Management* is the perfect revision aid for students preparing to take their NEBOSH Certificate in Environmental Management. As well as being a companion volume to Brian Waters' NEBOSH-endorsed textbook *Introduction to Environmental Management*, it will also serve as a useful aide-memoire for those in environmental management roles. The book aims to:

▶ Provide practical revision guidance and strategies for students
▶ Highlight the key information for each learning outcome of the current NEBOSH syllabus
▶ Give students opportunities to test their knowledge based on NEBOSH style questions and additional exercises
▶ Provide details of guidance documents publically available that students will be able to refer to.

The revision guide is fully aligned to the current NEBOSH syllabus, providing complete coverage in bite-sized chunks, helping students to learn and memorise the most important topics. Throughout the book, the guide refers back to the *Introduction to Environmental Management*, helping students to consolidate their learning.

Jonathan Backhouse is a chartered health and safety practitioner, qualified teacher and author. Since becoming self-employed in 2002 he has gained a vast amount of consultancy and training experience (working in the UK, Europe, Africa, the Middle East and the USA) within all aspects of health and safety, including fire safety, first aid, food hygiene and environmental management.

His qualifications include a Master of Arts in Professional Practice, in the context of Education, and a Master of Research in Occupational Health, Safety and Environmental Management.

Environmental Management Revision Guide

For the NEBOSH Certificate in Environmental Management

Jonathan Backhouse

Routledge
Taylor & Francis Group

LONDON AND NEW YORK

First published 2018
by Routledge
2 Park Square, Milton Park, Abingdon, Oxon OX14 4RN

and by Routledge
711 Third Avenue, New York, NY 10017

Routledge is an imprint of the Taylor & Francis Group, an informa business

© 2018 Jonathan Backhouse

Permission to reproduce extracts from British Standards is granted by BSI
Standards Limited (BSI). No other use of this material is permitted. British
Standards can be obtained in PDF or hard copy formats from the BSI
online shop: www.bsigroup.com/Shop.

British Library Cataloguing-in-Publication Data
A catalogue record for this book is available from the British Library

Library of Congress Cataloging-in-Publication Data
A catalog record for this book has been requested

ISBN: 978-0-415-79160-1 (hbk)
ISBN: 978-1-138-64912-5 (pbk)
ISBN: 978-1-315-62597-3 (ebk)

Typeset in Univers
by Apex CoVantage, LLC

Contents

Figures

Tables

Tables

Preface

Welcome to the first Revision Guide for the *NEBOSH Certificate in Environmental Management*.

The guide has been designed to be used together with the latest NEBOSH syllabus guide and the textbook *Introduction to Environmental Management: for the NEBOSH Certificate in Environmental Management*.

This guide gives a basic summary of the *NEBOSH Certificate in Environmental Management* course; a full explanation of all the topics is given in the relevant textbook.

The Revision Guide has the following features:

▶ It follows the latest NEBOSH (May 2012) syllabus.
▶ Revision notes are included for each of the elements.
▶ A summary of the learning outcomes and key points is given for each element.
▶ Important diagrams are included to help revision.

The Revision Guide will also be useful to those who have specific environmental responsibilities in their roles within the workplace and those who are studying other environmental qualifications.

This book has been written to help students prepare for their National Examination Board in Occupational Safety and Health (NEBOSH) Certificate in the Environmental Management exam. Introduction to the NEBOSH Environmental Management Certificate covers the fundamentals of Environmental Management for students who complete the NEBOSH National and International Certificate in Environmental Management.

This book covers the 2012 NEBOSH syllabus (publication date: April 2014). This first edition provides students with all they need to tackle the course with confidence. Whilst this book is a companion to the *Introduction to Environmental Management: for the NEBOSH Certificate in Environmental Management* course book, it can also be used as part of personal online study or in a classroom based course using other course books/material.

Whilst this is an international revision guide references have also been included for UK legislation. You will not be examined on UK legislation. You are expected to have a basic understanding of environmental legislation, in your own country, this will not be examinable, but you may refer to this in your exams. For example, stating the name of a key piece of legislation.

Institute of Environmental Management and Assessment (IEMA) have confirmed (in 2016), that those delegates holding the *NEBOSH Certificate in Environmental Management* may join their organisation and then use the post-nominal designated letters AIEMA (Associate membership).

All the best with your studies.

Jonathan Backhouse

MRes MA BA(Hons) DipNEBOSH
EnvDipNEBOSH CertEd CMIOSH

Acknowledgements

I would like to thank, Steve Goodgroves, Linda McGravey and Brian Waters for the support and technical expertise in writing this Revision Guide. I would also like to acknowledge the production team who worked on this book, Michael McKeen, indexer, Rodney Williams, copy editor, Allison Morgan, proofreader, and Kerry Boettcher, project manager. None of this would have been possible without the love, support and encouragement of my wife, Diane. Thank you!

EC 1.1

Foundations in environmental management

Learning outcomes	
1.1 Outline the scope and nature of environmental management	☐
1.2 Explain the ethical, legal and financial reasons for maintaining and promoting environmental management	☐
1.3 Outline the importance of sustainability and its relationship with corporate social responsibility	☐
1.4 Explain the role of national governments and international bodies in formulating a framework for the regulation of environmental management	☐

1.1 Outline the scope and nature of environmental management

> **Key revision points**
> ▶ Definition of the environment
> ▶ The multi-disciplinary nature of environmental management and the barriers to good standards of environmental management within an organisation
> ▶ The size of the environmental 'problem' in terms of the key environmental issues

Definition of the environment

The environment has been defined as

The surroundings in which an organisation operates whether they are man-made or natural, including air, water, land, natural resources, flora, fauna, humans and their interrelation.

'Surroundings' can extend from within an organisation to the global system.

The multi-disciplinary nature of environmental management

The multi-disciplinary nature of environmental management requires environmental managers to be able to address a wide range of factors, including quality, safety and health. For example, when purchasing new equipment the following environmental issues may need to be addressed

▶ Assessing the effects of any waste likely to be produced during the use of products or at the end of their life, their recyclability and any likely regulatory requirements.
▶ Atmospheric emissions arising from the use of the products/equipment.

▶ Consideration of the need to undertake life cycle assessments and or circular economy of products where appropriate and vet the supply chain.

▶ Effluents produced during use and their effects.

▶ Energy used during use of the products/equipment.

▶ Hazardous properties associated with the substances and degradation products, including such properties as flammability, toxicity, carcinogenicity, etc.

▶ Licensing or authorisation requirements for use or storage of substances.

▶ Packaging required and its disposal.

▶ Potential to cause nuisance through noise or smell.

▶ Restricting or avoiding products or equipment containing hazardous substances – mercury, cadmium, persistent chlorinated hydrocarbons, pesticides, etc.

▶ Transport impacts, such as emissions and noise.

The size of the environmental 'problem' in terms of environmental issues includes:

▶ local effects of pollution (noise, waste, lighting, odour);

▶ carbon emissions and climate change;

▶ air pollution and the ozone layer;

▶ water resources and pollution (nitrates);

▶ deforestation, soil erosion and land quality;

▶ material resources and land despoliation (land rights, etc.);

▶ energy supplies;

▶ waste disposal and international waste trade;

▶ agricultural issues arising from trade between developing and developed economies (e.g., create landfill sites in place of agricultural);

▶ desertification; and

▶ cultural heritage and material assets.

These are viewed as man-made.

Examples of man-made effects on the environment include:

(a) **Atmospheric pollution** – from vehicle traffic and local industrial processes, etc.

(b) **Aquatic contamination** – from accidental spillages to deliberate pollution and from fire-water

(c) **Land contamination** – from local industrial processes, for example, spills of chemicals and metals, and waste disposal, leachate from landfill, intense farming utilisation of pesticides and nitrates

(d) **Effects on the community** – caused by odour, noise, visual impact, lighting glare or any other effluvia (nuisance) and loss of amenity

(e) **Effects on the ecosystem** – loss of biodiversity and interdependent ecosystems

(f) **Loss of raw materials and natural resources/resource depletion** – loss of natural resources (which include non-renewable resources – for example, minerals and oil – and renewable resources, such as timber)

Key environmental issues to be addressed with regard to environmental management include:

▶ global warming;
▶ carbon emissions and climate change;
▶ air pollution and the ozone layer;
▶ water resources (potable) and water pollution;
▶ waste disposal and international waste trade;
▶ energy supplies;
▶ local effects of pollution (e.g., noise, waste, lighting, odour);
▶ agricultural issues arising from trade between developing and developed economies;
▶ deforestation, desertification, soil erosion and land quality; and
▶ material resources and land despoliation.

There are different cycles that are used within the environment, including:

▶ carbon (see next section);
▶ nitrogen (Chapter 3); and
▶ water (Chapter 5).

Carbon cycle

The carbon cycle is the process in which carbon travels from the atmosphere into organisms and the earth, then back into the atmosphere. The stages in the carbon cycle are respiration, photosynthesis, decomposition and combustion.

▶ **Respiration** by animals and plants releases energy. This means carbon is produced.
▶ **Photosynthesis** by plants creates energy. This means carbon is used.
▶ **Decomposition** occurs when an animal dies, which releases carbon back into the atmosphere.
▶ **Combustion** is burning; if something with carbon is burnt it will release it into the atmosphere, e.g., fossil fuel.

Carbon is stored in:

▶ the atmosphere;
▶ oceans;
▶ soil and rocks; and
▶ dead plants/animals (i.e., fossils).

Revision exercise

Outline possible benefits to an organisation for having good environmental management systems in place. (8)

Exam tip

The answers to the above question could include:

1 Cost savings by reducing the use of water, energy, raw materials and packaging
2 More efficient use of transport
3 Reduction of waste going to landfill and increase recycling
4 Not paying for waste transfer permits
5 Improved public relations (PR)
6 Improved legal compliance
7 Improved environmental performance by reducing risk
8 Improved relationships with internal/external stakeholders

9 Improved staff morale and recruitment
10 Positive view by insurers and investors
11 Better opportunities for marketing and other business development
12 Opportunities to draw down funding from governmental bodies

1.2 Explain the ethical, legal and financial reasons for maintaining and promoting environmental management

Key revision points

▶ The rights and expectations of local residents, including indigenous peoples, supply chain, customers and employees
▶ Outcomes of incidents in terms of environmental and human harm, and legal and economic effects on the organisation
▶ The actions and implications of pressure groups
▶ Overview of legal issues – breaches of national or local laws and individual legal rights
▶ Penalties such as fines/imprisonment and rights to compensation
▶ Different levels of standards and enforcement in many jurisdictions; the role of responsible business
▶ The business case for environmental management: direct and indirect costs of environmental accidents: insured and uninsured costs

Introduction to ethical, legal and financial reasons

The ethical, legal and financial reasons for maintaining and promoting environmental management include:

Ethical/ Moral: We all have an ethical or moral duty to be good stewards of the earth. A famous quote sums up this philosophy: 'We do not inherit the earth from our ancestors; we borrow it from our children.' We all have a general duty of care, which comes from our society's attitude to environmental issues seeing the need for sustainable development. What is important to remember is that sustainable development is a process that will allow for human and environmental relationships to exist sustainably.

Legal/ Social: Many countries have a large and growing amount of statutory environmental legislation and civil judgments. In addition there have been many international treaties; for example, Agenda 21. Ensuring compliance will help to reduce the risk of:

- ▶ effects of preventative measures (enforcement notices, permits, etc.);
- ▶ punitive measures through criminal sanctions; and
- ▶ compensatory effects of environmental law.

Financial/ Economic: Possibly the strongest argument for promoting environmental management is an economic one. The business case will involve identifying ways to reduce costs and fines, plus the ability to increase profitability. This will help reduce:

- ▶ direct costs associated with environmental pollution;
- ▶ indirect costs associated with environmental pollution;
- ▶ environmental taxation;
- ▶ tax relief schemes.

Additional financial benefits:

- ▶ cost savings;
- ▶ business benefits such as green claims;
- ▶ grants and funding.

The rights and expectations of local residents, including indigenous peoples

Indigenous peoples are custodians of some of the most biologically diverse territories in the world. They are also responsible for a great deal of the world's linguistic and cultural diversity, and their traditional knowledge has been, and continues to be, an invaluable resource that benefits all of mankind. Yet, indigenous peoples continue to suffer discrimination, marginalisation, extreme poverty and conflict. Some are being dispossessed of their traditional lands as their livelihoods are being undermined. Meanwhile, their belief systems, cultures, languages and ways of life continue to be threatened, sometimes even by extinction.

Environmental Supply Chain Management (ESCM) or Sustainable Supply Chain Management (SSCM) has been developed in recent years to help organisations address environmental management issues.

Supply Chain Management will help reduce costs by:

▶ improving efficiency;
▶ reducing demand for materials;
▶ cutting wastes;
▶ securing existing and future contracts from customers;
▶ reducing business risks;
▶ improving employee morale and working conditions; and
▶ using a model such as 'porters value chain' that will help to identify quantifiable inputs/outputs for goods and services.

Outcomes of incidents in terms of environmental and human harm, and legal and economic effects on the organisation

There have been countless man-made and natural incidents that have caused harm to both humans and the environment. These include nuclear, chemical and oil-based incidents.

Nuclear

The latest nuclear incident struck Japan's Fukushima Daiichi (2011) causing the most extensive release of radioactivity since the Chernobyl (1986) incident.

Chemical/oil

The latest major environmental disaster was the Deepwater Horizon oil spill (2010). Other major environmental disasters include Buncefield (2005), Exxon Valdez (1989), Bhopal (1984), Seveso (1976) and the Minamata Disaster (1956–59).

The actions and implications of pressure groups

Environmental pressure groups are organisations set up to try to influence the way in which someone thinks about the environment.

Study point

You will not need to know the specific details of the nuclear, chemical, or oil-based disasters for the exam.

Examples of environmental pressure groups include:

▶ Campaign for Better Transport;
▶ Campaign to Protect Rural England;
▶ Campaign for the Protection of Rural Wales;
▶ Greenpeace;
▶ Friends of the Earth (England, Wales and Northern Ireland);
▶ Friends of the Earth Scotland;
▶ Forum for the Future;
▶ Surfers Against Sewage;
▶ Rowers Against Thames Sewage (RATS);
▶ Waste Watch;
▶ Rights of Way Alliance Neath (Rowan); and
▶ Women's Environmental Network.

Currently, there is mixed public opinion towards Hydraulic Fracturing (fracking). This is a technique in which typically water is mixed with sand and chemicals that are then injected at high pressure into a wellbore to create fractures, which form conduits along which fluids

9

such as gas, petroleum and groundwater may migrate to the well. In the UK a pressure group specifically addressing fracking is Frack Off (http://frack-off.org.uk).

Overview of legal issues – breaches of national or local laws and individual legal rights

Many countries have a large and growing amount of statutory environmental legislation and civil judgments. In addition, there have been many international treaties.

Within international law there are two types of arrangements:

Hard law: Legally binding – directly enforceable by a national/international body. These are set within: Directives, Convections, Protocols and Treaties.

Soft law: Not legally binding – not enforceable by a national/international body. These are set within: Agendas, Charters and Recommendations.

Study point

The following legal issues are referenced for information, applying to the UK and other countries, and do not form part of the NEBOSH exam.

For a tort of negligence (which includes acts or omissions) to apply, it would need to be shown that:

► a duty of care was owed;
► this duty was breached;
► as a result of the breach harm or loss occurred; and
► the harm was foreseeable and not too remote from the breach.

A tort of negligence may be private or public.

Private nuisance is basically an unreasonable interference with a person's use or enjoyment of land, or some right over, or in connection with it. To be liable under the tort, it should be foreseeable that actions would be likely to give rise to a nuisance. Typical activities actionable under private nuisance include:

- ▶ encroachment (e.g., landslide);
- ▶ physical damage to land (e.g., migrating landfill gases killing vegetation); or
- ▶ interference with enjoyment of property (e.g., noise or smells).

Private nuisance is actionable by individuals with a direct proprietary interest in the land in question. Liability for an unreasonable interference or nuisance depends on a range of factors, such as:

- ▶ duration of the interference;
- ▶ the sensitivity of the claimant;
- ▶ any malice; and
- ▶ the character of the neighbourhood.

Public nuisance is similar to private nuisance, except that it is well established that there is no need to have an interest in land affected and the prescription is not a defence. Persons affected are the public, or a section of it, which suffer damage at large. Typical examples of public nuisance would be:

- ▶ wide-scale fallout of dust over a large number of properties;
- ▶ blasting noise and flying rocks from a quarry; or
- ▶ an offensive smell affecting a town centre.

Civil law and criminal law are different **types** of law.

- ▶ **Civil law** is mainly derived from common law.
  - ▷ Civil law is concerned with the rights and duties of individuals and organisations towards each other.
  - ▷ Violations are known as torts.
  - ▷ Remedies may be in the form of damages or compensation.
  - ▷ The standard of proof is the 'balance of probabilities'.
- ▶ **Criminal law** is concerned with breaches of statute.
  - ▷ Statutory bodies such as the environment agency or local authorities bring prosecutions.
  - ▷ The main aim of prosecution is to punish, deter and issue sanctions: these can be fines or imprisonment.
  - ▷ The standard of proof is, therefore, 'beyond reasonable doubt'.

Common law and statute law are primary **sources** of law.

▶ **Common law** is produced from published judicial opinions (also known as case law)
  ▷ Accumulated judicial decisions made by courts when hearing similar cases.
▶ **Statute law** is produced by Parliament and written in statutes or Acts of Parliament and
  ▷ Acts supersede all other forms of law.
  ▷ Only Parliament can make, modify, amend or revoke statutes.

For example in the UK, legislation is extensive. It includes over 100 regulations for:

▶ air pollution;
▶ waste management;
▶ product and packaging;
▶ water, groundwater and land contamination;
▶ hazardous materials;
▶ community, grounds and biodiversity; and
▶ construction and buildings.

Penalties such as fines/imprisonment and rights to compensation

Many countries have their own enforcement bodies. For example, in England and Northern Ireland the Environment Agency (EA), in Wales the Natural Resource Wales and in Scotland the Scottish Environmental Protection Agency (SEPA) take on this role.

Courts can:

▶ impose fines;
▶ impose custodial sentences;
▶ require payments for claims for damages;
▶ require a company to pay for clean-up costs to rectify environmental damage;
▶ require a company to carry out works to prevent future pollution; and
▶ serve sanctions/suspension of operating licenses.

Environmental law across the world is varied and vastly complicated; however, the key principles that apply are:

▶ the producer/manufacturer needs to consider their impacts on the environment;
▶ to ensure that pollution prevention in one area does not lead to the pollution in another, an integrated pollution prevention and control (IPPC) methodology should be implemented;
▶ a preventative/precautionary principle should apply towards dealing with the environment;
▶ preventative approach;
▶ producers' responsibilities;
▶ life cycle analysis;
▶ best available techniques; and finally,
▶ the polluter pays!

Environmental offences may include:

▶ breaches of permits, licences and consents;
▶ illegal discharges to air, land and water;
▶ transportation of controlled waste without registering;
▶ breach of duty of care; and
▶ breach of an abatement notice.

Different levels of standards and enforcement in many jurisdictions; the role of responsible business

It is possible for a national government to take legal action against an organisation in the event of environmental pollution occurring (criminal law). This can be on an international scale; for example, in the case of the Deepwater Horizon oil spill, the claims and fines are estimated to be over $45 billion. The penalties for environmental offences may include fines, legal costs, cleanup costs (restitution remedies), environmental costs and in some cases imprisonment.

A civil action generally involves individuals: a claimant suing a defendant for a remedy or remedies. These are often for statutory nuisances (i.e., emissions of smoke, gases, fumes, dust, noise, odour, build up of rubbish, non-natural light and animals).

Within the UK and other countries, enforcement notices that may be served by an environmental inspector could include:

▶ **Prohibition notices** where there is imminent risk of serious environmental damage from an activity not covered by an environmental permit.

▶ **Suspension notices** where there is serious risk of pollution from an authorised process.

▶ **Variation notices** if the conditions of an authorisation require being changed.

▶ **Revocation** of part or all of a permit if the person operating an installation ceases to be a fit and proper person or ceases to be the operator.

▶ **Enforcement notice** if permit conditions are not being met.

▶ **Abatement notice** by a local authority for causing nuisance.

▶ **Remediation notice** in respect of contaminated land.

Study point

The specific enforcement notices are referenced for information, applying to the UK and other countries, and do not form part of the NEBOSH exam.

Within the UK, defences that can be used in an environmental case include:

1 **Breach by a third party** – The defendant cannot be held responsible if the defendant did not commit it.

2 **Consent of the claimant** – The defendant cannot be held responsible if the claimant has agreed to the nuisance.

3 **Lack of foreseeability** – The defendant cannot be held responsible if the circumstances are unforeseeable of an event or action happening.

4 **Necessity** – The defendant may not be held responsible if it can be proven that the action was of 'necessity' or 'no reasonable alternative' was possible.

5 **No breach of duty** – The defendant cannot be held responsible if there is no breach of duty.

6 **No duty owed** – The defendant cannot be held responsible if there is no duty of care owed.

7 **Prescription** – The defendant may not be held responsible if it can be shown that nuisance has been carried out for a long time without complaint from the claimant.

8 **Remoteness of damage** – The defendant may not be held responsible if it can be proven that the damage or harm was not reasonably foreseeable.

9 **Statutory authority** – The defendant may not be held responsible if the actions that cause harm/nuisance are being permitted in the exercise of a statutory power.

10 **The breach did not lead to the damage** – The defendant may not be held responsible if the claimant has not suffered a harm/nuisance.

11 **Volenti non fit injuria** – 'that no injury can be done to the willing'. The defendant may not be held responsible if the claimant permitted the harm/nuisance.

Study point

The above defences are referenced for information, applying to the UK, and do not form part of the NEBOSH exam.

You will not need to know the specific details; marks may be available for knowing that a range of enforcement notices are available for enforcement officers and a range of defences are available for companies.

The business case for environmental management: direct and indirect costs of environmental accidents: insured and uninsured costs

The costs that are related to environmental accidents include:

Insured direct costs

▶ claims on employers' and public liability insurance;

▶ damage to buildings or vehicles; and
▶ damage to people, land, plants or animals.

Insured indirect costs

▶ public and third part liability.

Uninsured direct costs

▶ fines and clean up costs;
▶ increases in insurance premiums resulting from the accident;
▶ any compensation not covered by the insurance policy due to excess agreed between the employer and the insurance company; and
▶ legal representation.

Uninsured indirect costs

▶ loss of goodwill and a poor corporate image;
▶ the accident investigation time; and
▶ production delays.

Revision exercise

Identify direct and indirect costs that may be related to environmental accidents.

Exam tip

Differences between identify and outline often confuse students with exam questions.

The above section on *defences* shows what would be required for both identify and outline; i.e., if the question was to **identify** types of defences and **outline** their meaning. The identify part is in bold; the rest is sufficient for an outline.

1.3 Outline the importance of sustainability and its relationship with corporate social responsibility

Key revision points

▶ Meaning of sustainability (with reference to Rio Earth summit)
▶ Importance of sustainable development

Meaning of sustainability

Sustainability has been defined as:

The right to development must be fulfilled so as to equitably meet developmental and environmental needs of present and future generations.

Other definitions include:

Meeting the needs of current generations without compromising future generations.

Concept of balancing social, financial (economic) and environmental considerations.

Sustainable development is development that meets the needs of the present without compromising the ability of future generations to meet their own needs. It contains within it two key concepts (Our Common Future – Brundtland, 1987):

▶ The concept of 'needs', in particular, the essential needs of the world's poor, to which overriding priority should be given.
▶ The idea of limitations imposed by the state of technology and social organisation on the environment's ability to meet present and future needs.

Sustainable development can be achieved by means of:

- ▶ effective protection of the environment;
- ▶ prudent use of natural resources;
- ▶ economic development and maintenance of stable levels of growth;
- ▶ relationship of environmental performance and sustainability to corporate social responsibility and environmental risks from outsourcing; and
- ▶ social progress.

Importance of sustainable development

The Rio Declaration on Environment and Development and the Statement of Principles for the Sustainable Management of Forests were adopted by more than 178 governments at the United Nations Conference on Environment and Development (UNCED) held in Rio de Janerio, Brazil, 3 to 14 June 1992. The '21' in Agenda 21 refers to the 21st century. It was a non-binding, voluntarily implemented action plan of the United Nations with regard to sustainable development.

In 2012 the third international conference on sustainable development aimed at reconciling the economic and environmental goals of the global community became known as Rio 2012, **Rio+20** or Earth Summit 2012. Rio+20 has highlighted seven areas that need priority attention:

1   decent jobs;
2   energy;
3   sustainable cities;
4   food security;
5   sustainable agriculture;
6   water and oceans; and
7   disaster readiness.

Revision exercise

(a) **Give** the meaning of sustainability. (2)
(b) **Give** the meaning of the environment. (2)
(c) **Outline** how sustainable development can be achieved by a business. (2)
(d) **Identify** examples of actions that businesses could take to demonstrate sustainability. (2)

Exam tip

Creating a list of keywords and their meanings will help with your revision.

1.4 Explain the role of national governments and international bodies in formulating a framework for the regulation of environmental management

Key revision points

▶ International law governing the environment (e.g., OSPAR Convention, Montreal Protocol, Basel Convention, Ramsar Convention)
▶ The role of the European Union in harmonising environmental standards
▶ The importance of knowing and understanding local legislation
▶ Meaning of Best Available Technique (BAT) and Best Practicable Environmental Option (BPEO)
▶ The role of enforcement agencies and the consequences of non-compliance

International law governing the environment

Convention for the Protection of the Marine Environment of the North-East Atlantic 1992 (the 'OSPAR Convention')

OSPAR is the mechanism by which fifteen governments of the western coasts and catchments of Europe, together with the European Union, cooperate to protect the marine environment of the North-East Atlantic.

Protocol on Substances that Deplete the Ozone Layer 1987 (the Montreal Protocol/Vienna Convention)

Sets targets for reducing and eliminating the production and consumption of ozone-depleting substances, makes financial provision for developing countries in terms of alternatives to ozone-depleting chemicals and bans trade of ozone-depleting substances with non-signatory parties. For example, the ban on Halon fire extinguishers was implemented following the Montreal Protocol and subsequent extension at Kyoto a decade later.

Convention on the Control of Transboundary Movements of Hazardous Wastes and their Disposal 1989 (the Basel Convention)

To control the export of hazardous/toxic waste and dumping by industrial nations in developing countries, and to reduce the amount of hazardous waste generated.

Convention on Wetlands 1989 (the Ramsar [Iran] 1971 Convention)

This is an intergovernmental treaty which provides the framework for conservation and wise use of wetlands and their resources. All the sites are protected under wildlife legislation (through their notification as Sites of Special Scientific Interest).

COP 21

The 21st Session of the Conference of the Parties to the United Nations Framework Convention on Climate Change (COP 21).

The two-week Paris climate change summit (December 2015) set the aim to keep a global temperature rise this century well below 2 degrees Celsius and to drive efforts to limit the temperature increase even further to 1.5 degrees Celsius above pre-industrial levels.

The role of the European Union in harmonising environmental standards

Since Brexit, the details of how the role of the European Union's environmental standards will impact the UK in the future is unknown. **Changes will not impact the syllabus or questions for at least six months.**

Since the 1970s, the EU has agreed to over 200 pieces of legislation to protect the environment. Member States are responsible for implementing EU environmental legislation. Every country that applies to join the Union has to harmonise its environmental norms with those of the EU. The European Commission can, through the European Court of Justice, take legal action against a Member State that fails to implement legislation correctly.

Within Europe, the majority of the environmental legal requirements for the Member States originate from the EU in the form of regulations or directives.

EU Regulations become law without requiring further action from the Member State, whereas the EU Directives require the Member State to implement these at their discretion.

The European Environment Agency (EEA) is an agency of the European Union. The aim of the EEA is to ensure that decision-makers and the general public are kept informed about the state and outlook of the environment.

EEA's mandate is:

▶ To help the community and member countries make informed decisions about improving the environment, integrating environmental considerations into economic policies and moving towards sustainability.
▶ To coordinate the European environment information and observation network.

The importance of knowing and understanding local legislation

Each country will, in addition to being required to comply with International Environmental Protocols and Conventions (and EU legislation if a Member State), need to comply with the local (national) legislation.

Meaning of Best Available Technique (BAT) and Best Practicable Environmental Option (BPEO)

The concept of Best Available Technique (BAT) means the most effective and advanced stage in the development of activities and their methods of operation which indicates the practical suitability of particular techniques for providing the basis for emission limit values and other permit conditions designed to prevent the release of emissions and where that is not practicable, to reduce emissions and the impact on the environment as a whole.

▶ **Best** means most effective in achieving a high general level of protection of the environment as a whole.
▶ **Available** techniques means those developed on a scale which allow implementation in the relevant industrial sector, under economically and technically viable conditions, taking into consideration the costs and advantages, whether or not

the techniques are used or produced inside the Member State in question, as long as they are reasonably accessible to the operator.

▶ **Techniques** include both the technology used and the way in which the installation is designed, built, maintained, operated and decommissioned.

Best Available Technique simply means that the operator has to use the very best possible way to protect the environment that can be economically justified.

The Royal Commission on Environmental Pollution proposed the concept of Best Practicable Environmental Option (BPEO) as a means of controlling pollution in 1988 and it is defined as:

The option which provides the most benefit or least damage to the environment as a whole, at acceptable cost, in the long term as well as the short term.

The role of enforcement agencies and the consequences of non-compliance

Typical powers available to an environmental inspector will vary from country to country. These may include:

▶ To enter premises at any reasonable time or at any time when it is considered that there is an immediate risk of serious pollution of the environment.

▶ Direct that all or part of the premises, or anything in them, be left undisturbed as long as is reasonably necessary for the purpose of any investigation.

▶ Take samples of any article or substance found.

▶ Dismantle or test any article or substance found.

▶ Take possession of any article or substance found and retain it.

▶ Require any person
  ▷ To answer such questions as the inspector thinks fit to ask; and
  ▷ To sign a declaration of the truth of their answers.

▶ Take witness statements.

▶ Take a copy of any record that is required to be kept.

▶ If they have reasonable cause to believe that an article or substance is the cause of immediate danger or serious harm and therefore seize it and cause it to be rendered harmless.

▶ Be accompanied by a police officer should obstruction be likely/prevent access.

▶ Make any investigation as necessary including measurements, taking samples and photographs and questioning individuals.

▶ Carry out experimental borings.

▶ Install and maintain monitoring equipment.

▶ In cases of emergency, gain entry at any time, with force if necessary.

▶ Start the prosecution process.

Other countries may have similar systems in place.

The consequences of non-compliance may result in:

▶ Criminal prosecutions by the enforcing authorities or through private prosecutions.

▶ Significant fines that can be given for serious breaches of the law.

▶ Potential for prison sentences and for personal prosecution for directors and managers.

▶ Administrative sanctions including serving notices such as revocation notices, suspension notices and enforcement notices.

▶ Orders requiring works to be undertaken to prevent pollution.

▶ Award of clean-up costs for rectifying environmental damage.

▶ Civil liability resulting in claims for damages or injunctions.

Revision exercise

Explain the key purposes for:

(a) Convention for the Protection of the Marine Environment of the North-East Atlantic 1992 (the 'OSPAR Convention'); (2)

(b) Protocol on Substances that Deplete the Ozone Layer 1987 (the Montreal Protocol); (2)

(c) Convention on the Control of Transboundary Movements of Hazardous Wastes and their Disposal 1989 (the Basel Convention); (2)

(d) Convention on Wetlands 1989 (the Ramsar 1971 Convention). (2)

Exam tip

Specific examples of local (national) legislation and case law are not required for this qualification.

EC 1.2

Environmental management systems

Learning outcomes

2.1 Identify the reasons for implementing an environmental management system (EMS)	☐
2.2 Describe the key features and appropriate content of an effective EMS, i.e., ISO 14001	☐
2.3 Outline the benefits and limitations of introducing a formal EMS such as ISO 14001/BS 8555/EMAS into the workplace	☐
2.4 Identify key members of the ISO 14000 family of standards and their purpose	☐

2.1 Identify the reasons for implementing an environmental management system (EMS)

Sharing of common management system principles with quality and health and safety management – enabling integration

Learning outcomes

▶ ISO 9001 Quality Management;
▶ ISO 45001 Health and Safety Management; and
▶ ISO 14001 Environmental Certification.

The benefits and limitations to an organisation of integrating its currently separate systems for health and safety management, quality and environmental management include:

Benefits

▶ Avoiding duplication of information in areas of overlap (e.g., hazardous substances control, risk assessment, storage standards).
▶ Control systems could be designed to minimise conflicts, such as design of ventilation systems transferring pollutants from the workplace to the outside atmosphere.
▶ Decreased time is taken for decision-making.
▶ Documented control systems that cover both disciplines would avoid additional paperwork.
▶ Integrated training.
▶ Integration of emergency responses.

- Potential cost savings.
- Potential for combined auditing.
- Potential for further integration with other areas of management control, such as quality.
- Reduced administration.
- Systems that work to maximise benefit for the organisation, rather than pulling against each other.
- Better relationships with internal/external stakeholders.
- Better supply chain management.
- Potential increase in business profits.

Limitations

- Harder for the regulator to identify relevant aspects of the system.
- Larger, more complex systems may be more difficult to change and slower to respond to pressure for change.
- More complex certification process where external certification was sought.
- More complex documentation systems.
- Possible reduction in sense of ownership of the system.
- Potential for role conflicts between managers.
- Increased cost.
- Ongoing management issues.
- Increased training costs for continuous improvement.
- Costs may well exceed the business benefits.
- Ongoing commitment.

Addressing stakeholders' views

Stakeholders can be internal or external to an organisation. The internal stakeholders are directors, trustees, workers, etc., whereas the list of external stakeholders is more extensive and generally includes anyone who has an interest in the organisation including:

- investors;
- customers and suppliers;
- insurance companies;

- ▶ regulators, e.g., environmental bodies set up by national governments;
- ▶ neighbours (those who are directly/indirectly affected by the organisation's activities);
- ▶ parent companies;
- ▶ financial institutions;
- ▶ directors;
- ▶ non-government agencies; and
- ▶ consumers.

Stakeholders are those groups and individuals with an interest in an organisation/ business/activity.

Corporate Social Responsibility

Increasingly, stakeholders (defined as those who have an interest in the organisation) expect that organisations should be more environmentally and socially responsible in conducting their business. One approach to this is through Corporate Social Responsibility (CSR). CSR is a form of corporate self-regulation integrated into a business model.

There's no single definition of CSR as every organisation, government and individual have a different angle.

> [CSR] is the continuing commitment by business to behave ethically and contribute to economic development while improving the quality of life of the workforce and their families as well as of the local community and society at large.

The Chartered Institute of Public Relations defines CSR as:

> A concept whereby companies integrate social and environmental concerns in their business operations and in their interaction with their stakeholders on a voluntary basis.

Factors to be considered when developing a CSR policy will include how an organisation:

- ▶ Conducts its business – in the marketplace and through its supply chain.

▶ Makes strategic decisions (e.g., how it factors in environmental and social considerations).
▶ Manages and rewards its people through a performance management and remuneration process that takes a balanced view of an individual's performance – not just their financial performance.
▶ Reports its business performance.

The two key reasons why organisations develop CSR reports are:

▶ To show stakeholders that a company is investing in the community and environmental stewardship.
▶ To benchmark a company's current operations and impact on society and the environment in order to know how and where they can improve.

ISO 26000:2010 is the Guidance for Social Responsibility and the Earth Charter.

Revision exercise

Identify possible internal and external stakeholders of an international organisation. (8)

Exam tip

List internal/external stakeholders that have an impact on your organisation and outline what that impact is.

2.2 Describe the key features and appropriate content of an effective EMS, i.e., ISO 14001

Key revision points

▶ Initial environmental review
▶ Environmental policy

▶ Plan-Do-Check-Act
▶ Planning
▶ Implementation and operation
▶ Checking
▶ Environmental auditing
▶ Active monitoring measures
▶ Reactive monitoring measures
▶ Review of environmental performance
▶ Management review
▶ Continual improvement

Initial environmental review

Prior to developing the EMS an **initial environmental review** needs to be undertaken.

This will form the baseline and foundation for the other parts of an EMS. Factors to consider when developing an environmental policy will include a detailed examination of all

▶ details from emissions, discharges and material and utility use;
▶ environmental impacts (for normal operating conditions, abnormal conditions and emergency situations) that are significant and need improvement by setting objectives and targets;
▶ possible breaches or potential breaches of environmental legislation;
▶ possible opportunities for improving performance and minimising waste; and
▶ previous emergency situations and accidents.

Environmental policy

The **environmental policy** will be appropriate in nature and scale to the organisation's activities and demonstrate a commitment to continual improvement. This will help to ensure that the organisation complies with local legislation.

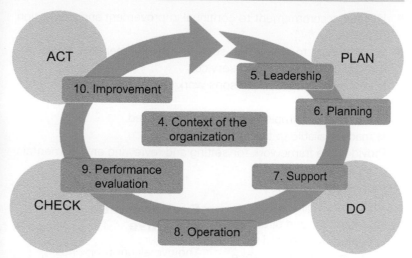

Figure 2.1 1400:2015

(Source: BSI Moving from ISO 14001:2004 to ISO 14001:2015)

ISO 14001 – Environmental Management Systems, is the world's leading international environmental standard. ISO 14001 has the new high level structure (HLS) that brings a common framework to all management systems.

The Plan–Do–Check–Act (PDCA) cycle can be applied to all processes and to the environmental management system as a whole. Figure 2.1 illustrates how Clauses 4 to 10 can be grouped in relation to PDCA.

The headings of Plan–Do–Check–Act have been used following the NEBOSH syllabus

Plan

The environmental policy statement that:

▶ Includes a commitment to comply with applicable legal requirements and with other requirements to which the organisation subscribes which relate to its environmental aspects.

▶ Includes a commitment to continual improvement and prevention of pollution.
▶ Is appropriate to the nature, scale and environmental impacts of its activities, products and services.
▶ Is communicated to all persons working for, or on behalf of, the organisation.
▶ Is documented, implemented and maintained.
▶ Is made available to the public.
▶ Provides the framework for setting and reviewing environmental objectives and targets.

ISO 14001:2015 defines environmental policy as the overall intentions and direction of an organisation related to its environmental performance as formally expressed by top management.

The **planning** stage of the EMS will encompass the development of specific targets and objectives.

The organisation should prepare written procedures for establishing and reviewing objectives and targets set by management.

In addition specific arrangements will need to be developed, for example, currently ISO 14001 requires an organisation to establish and maintain evaluation procedures for the determination of significant actual or potential environmental impacts.

Environmental objective

The overall environmental goal, consistent with the environmental policy, that an organisation sets itself to achieve. For example, reduce the amount of waste generated and its environmental impact.

Environmental target

Detailed performance requirement, applicable to the organisation or parts thereof, that arises from the environmental objectives and that needs to be set and met in order to achieve those objectives. For example, by 31 December 2018 reduce the waste going to landfill by 20 per cent.

The planning stage of 14001:2015 incorporates the following three concepts:

▶ environmental aspects;
▶ legal and other requirements; and
▶ objectives, targets and programme(s).

The organisation should consider their activities (aspects) and what impact these might have on the environment:

▶ emissions to air;
▶ releases to water;
▶ releases to land;
▶ use of raw materials and natural resources;
▶ use of energy;
▶ how energy is emitted, e.g., heat, radiation, vibration; and
▶ what and how waste and by-products are produced.

Potential impacts include:

▶ climate change; and
▶ health effects (lifestyle/ social/economic).

Aspect

Element of an organisation's activities or products or services that can interact with the environment; significant aspects will have a significant environmental impact.

Impact

Any change to the environment, whether adverse or beneficial, wholly or partially resulting from an organisation's environmental aspects.

Do

Implementation and operation

▶ Environmental risk profiling will include looking at the nature and level of risks faced by the organisation; the likelihood of any effects and their outcome. In addition, it will assess control measures – in essence, an environmental risk assessment. This will involve looking at the organisation's aspects and impacts.

- ▶ Part of the risk profiling should include the development of a legal register to ensure compliance with local legislation.
- ▶ Management needs to develop a plan that will manage the risks on the environment as a whole and at a local level, in addition to complying with legal and other standards.
- ▶ Environmental aspects and impacts, which will involve developing an Environmental Risk Assessment.
- ▶ Identify legal and other requirements, which will involve developing an Environmental Legal Register.

The implementation and operation stage of 14001 incorporates the following seven concepts:

- ▶ resources, roles, responsibility and authority;
- ▶ competence, training and awareness;
- ▶ communication;
- ▶ documentation;
- ▶ control of documents;
- ▶ operational control; and
- ▶ emergency preparedness and response.

An environmental awareness programme could be used to introduce the EMS to employees.

The programme would include:

- ▶ legal, financial and moral reasons for managing the environment;
- ▶ an overview of types and common causes of pollution;
- ▶ description of the organisation's operations and products;
- ▶ the organisation's environmental policy/environmental management system, etc.;
- ▶ the organisation's and the employees' legal responsibilities;
- ▶ how employees can play their part in meeting these obligations;
- ▶ waste minimisation;
- ▶ reporting problems;
- ▶ details on particular hazards;
- ▶ details on emergency plans associated with the site; and
- ▶ examples of previous incidents may be relevant.

Check

Checking and corrective action

▶ **Active monitoring**, for example:
  ▷ sampling of air quality;
  ▷ sampling of water emissions;
  ▷ boundary noise survey;
  ▷ mass balance calculations for fugitive releases to air;
  ▷ site inspections to identify potential risks;
  ▷ flow measurement of discharges to surface water; and
  ▷ the tonnage of waste generated each year.

▶ **Reactive monitoring**, for example:
  ▷ collecting data on near misses; and
  ▷ monitoring of complaints and comments from
    • workforce;
    • neighbours; and
    • enforcement actions.

▶ **Review of environmental performance**, including:
  ▷ incident data;
  ▷ inspections;
  ▷ control and monitoring of emissions;
  ▷ energy and raw material management;
  ▷ waste management;
  ▷ surveys, tours and sampling;
  ▷ quality assurance reports, audits, monitoring data/records/ reports, complaints;
  ▷ investigating environmental incidents and reporting requirements internally and externally;
  ▷ reporting on environmental performance;
  ▷ role of boards, chief executive/managing director and senior managers; and
  ▷ feeding into action and development plans as part of continuous improvement.

Environmental auditing is part of the checking process but is often looked at as a separate entity. An environmental audit is a systemic critical examination of an organisation's environmental management system, involving a structured process including the use of a series of questions and the examination of documentation.

The purpose is to collect independent information with the aim of assessing the effectiveness and reliability of the system and suggesting corrective action when this is thought to be necessary.

It is to be carried out by trained auditors, who may be internal or external to the organisation. In both cases, these types of audits can be seen as continuous or periodic monitoring.

An audit will look at specific issues including:

- ▶ environmental policy;
- ▶ effluent treatment and discharge;
- ▶ emissions to the atmosphere;
- ▶ incident and emergency response;
- ▶ land and property management;
- ▶ natural habitat;
- ▶ noise nuisance;
- ▶ production and operations;
- ▶ resources;
- ▶ storage of materials;
- ▶ training and systems of work;
- ▶ waste management;
- ▶ current legislation;
- ▶ current environmental practices;
- ▶ any previous audit findings;
- ▶ aspects and impacts; and
- ▶ any improvements using best practice.

The audit process forms three different aspects: preparatory work, on-site audit and conclusion work. Time spent can be calculated at a ratio of 20:60:20.

Preparatory work

- ▶ Meet with relevant managers and employee representatives to discuss and agree on the objectives and scope of the audit;
- ▶ prepare and agree on the audit procedure with managers; and

▶ gather and consider documentation.

On-site audit

▶ Interviewing;
▶ observations;
▶ review and assess additional documents; and
▶ observation of physical conditions and work activities.

Conclusion work

▶ Assemble the evidence;
▶ evaluate the evidence;
▶ write an audit report; and
▶ highlight any major or minor non-conformances.

Benefits of auditing are extensive. In addition to maintaining accreditation to 14001, it can help by:

▶ Checking compliance with legal requirements.
▶ Checking against internal targets.
▶ Checking adequacy of company policies and management systems.
▶ Checking the effectiveness of controls and procedures.
▶ Identifying areas for improvement.
▶ Raising the awareness of environmental issues to employees.
▶ Verifying credentials to external stakeholders.
▶ Prevention of environmental incidents.

Audits can be external and internal – both have advantages and associated disadvantages.

External audits can be advantageous because they:

▶ are usually impartial (auditor will have a range of experience of different types of work practices);
▶ may be able to offer solutions to what might be considered unsolvable;

▶ are not inhibited by criticism; and
▶ will assess the organisation's performance without prior bias.

However, external audit disadvantages include:

▶ need to plan well to identify the nature and scope of the organisation;
▶ individuals may not be forthcoming or be nervous or resistant to discussing their workplace with an outsider;
▶ may seek unrealistic targets; and
▶ cost more.

Internal audits can be advantageous because of:

▶ local acceptance to implement recommendations and actions;
▶ auditor's intimate knowledge of hazards and work practices;
▶ awareness of what might be appropriate for the industry;
▶ familiarity with the workforce, i.e., their strengths and weaknesses;
▶ relatively low cost and easier to arrange; and
▶ easy to plan.

However, internal audit disadvantages include:

▶ The auditor may not possess the correct auditing skills.
▶ The auditor may not be up-to-date with current legislation and best practices.
▶ The auditor may also be responsible for implementation of any proposed changes, and this might inhibit recommendations because of the effect on his workload.
▶ The auditor may be subject to pressure from management and time constraints.
▶ The auditor may be inhibited by bias.

Documents to be reviewed during an audit include:

- previous audits;
- records of
  - emission monitoring;
  - energy use;
  - raw materials use;
  - waste production; and
  - recycling and reuse;
- complaint data;
- enforcement notices and/or other communications from regulators;
- accident and incident reports;
- company policy and procedures and/or EMS manual;
- legal aspects register;
- consents and permits and evidence of compliance;
- maintenance logs;
- results of site inspections;
- staff training records;
- results of attitude surveys;
- minutes from management review meetings;
- emergency procedures; and
- customer feedback.

Act

Management review

14001 includes guidance on management review.

Selected members of the organisation's senior management team should carry out periodical reviews of the Environmental Management System. The following are the main inputs of the management review:

- Results of internal audits and evaluations of compliance with legal and other requirements.

▶ Communication from interested external parties, including complaints.
▶ Environmental performance of the organisation.
▶ The extent to which objectives and targets have been met.
▶ Status of preventive and corrective actions and who is responsible for them.
▶ Follow-up actions from previous management reviews.
▶ Changing circumstances, including developments in legal and other requirements.
▶ Recommendations for improvement.

Continual improvement

ISO 14001 and BS8555 both require continual improvement in an organisation's environmental performance. ISO 14001 defines continual improvement as:

A recurring process of enhancing the environmental management system in order to achieve improvements in overall environmental performance consistent with the organisation's environmental policy.

Revision exercise

(a) **Outline** reasons for producing a report on environmental performance. (6)
(b) **Identify** information that can be gathered to review environmental performance. (10)
(c) **Identify FOUR** different types of external reports containing environmental performance. (4)

Exam tip

Review your own organisation's EMS. What are the areas for improvement?

2.3 Outline the benefits and limitations of introducing a formal EMS such as ISO 14001/BS 8555/EMAS into the workplace

---

Key revision points

▶ Benefits of introducing ISO 14001/BS 8555/EMAS into an organisation

▶ Limitations of introducing ISO 14001/BS 8555/EMAS into an organisation

---

Benefits of introducing ISO 14001/BS 8555/EMAS into an organisation include:

▶ Increased compliance with legislative requirements.
▶ Competitive edge over non-certified businesses.
▶ Improved management of environmental risk.
▶ Increased credibility that comes from independent assessment.
▶ Savings from reduced noncompliance with environmental regulations.
▶ Heightened employee, shareholder and supply chain satisfaction and morale.
▶ Meeting modern environmental ethics.
▶ Streamlining and reducing environmental assessments and audits.
▶ Increased resource productivity.

Limitations of introducing ISO 14001/BS 8555/EMAS into an organisation include:

▶ Prescriptive environmental performance levels not included within the standard.
▶ Negligible improvements in environmental performance.
▶ Lack of public reporting for ISO 14001 unlike others internationally recognised.

▶ Management systems, i.e., Eco-Management and Audit Scheme (EMAS).

▶ Inconsistency of external auditors (e.g., differing levels of knowledge and the experience of the particular industry).

▶ Implementing an EMS may have costs that are too high for small and medium-sized enterprises.

Additional information for learning outcome

The three common EMSs are ISO 14001, Eco-Management Audit Scheme (EMAS) and BS 8555.

1 **ISO 14001** is the world's most popular standard for environmental management. Certification to ISO 14001:2004 will be allowed for a period of time following the publication of the new 2015 version of the standard, determined by the UK Accreditation Service (UKAS). Over 250,000 organisations are certified to ISO 14001, and while it continues to be as relevant as ever, the revision will take into consideration a number of issues to ensure that organisations are able to grow in a sustainable way.

2 **EMAS** is a voluntary initiative designed to improve companies' environmental performance. It is the premium environmental management tool to achieve this. It leads to enhanced performance, credibility and transparency of registered organisations.

3 **BS 8555** (full title: Guide to the phased implementation of an environmental management system including the use of environmental performance evaluation) is a standard that links Environmental Management Systems (ISO 14001) and Environmental Performance Evaluation (ISO 14031) to provide for focused training, auditing and implementation at each level and to support relationships between suppliers and customers.

The implementation of a general environmental management system (EMS) has many benefits, including:

▶ Able to set targets against which performance can be measured.

▶ EMS provides an audit trail.

- Improved (more efficient) use of transport.
- Improved environmental performance by reducing risk.
- Improved legal compliance.
- Improved management of environmental risk.
- Improved opportunities for competitive advantage against competitors without an EMS.
- Improved public relations (PR).
- Improved relationships with stakeholders.
- Improved staff morale and recruitment.
- Integration with safety and quality management systems offer cost savings and deliver other business benefits.
- Reduced environmental impacts by using fewer resources.
- Reduced insurance premiums.
- Reduced risks of prosecution.
- Reduced use of water, energy, raw materials and packaging.

In addition, a circular economy is a key benefit – which can be defined (by Wrap) as:

> An alternative to a traditional linear economy (make, use, dispose of) in which we keep resources in use for as long as possible, extract the maximum value from them whilst in use, then recover and regenerate products and materials at the end of each service life.

Limitations resulting from introducing an EMS include:

- Time and financial commitments both to introduce and then audit and maintain an EMS may be difficult, especially for a small business.
- Lack of management support.
- Resistance from staff.
- Lack of knowledge or skills required for implementation.
- No requirement for public reporting.
- In some cases, there may be negligible improvement in environmental performance.

Revision exercise

Identify **FOUR** benefits **AND FOUR** limitations of introducing an environmental management system based on BS 8555 and EMAS ISO 14001 into the workplace. **(8)**

Exam tip

Review your own organisation's EMS. What are the specific benefits and limitations of your organisation for having an EMS?

2.4 Identify key members of the ISO 14000 family of standards and their purpose

The ISO 14000 family addresses various aspects of environmental management. It provides practical tools for companies and organisations looking to identify and control their environmental impact and constantly improve their environmental performance. ISO 14001 and ISO 14004 focus on environmental management systems. The other standards in the family focus on specific environmental aspects such as life cycle analysis, communication and auditing.

▶ ISO 14004:2015, Environmental Management Systems – General guidelines on principles, systems and supporting techniques
▶ ISO 14006:2011, Environmental Management Systems – Guidelines for incorporating ecodesign
▶ ISO 14031:2013, Environmental Management – Environmental performance evaluation guidelines
▶ ISO 14044:2006, Environmental management – Life cycle assessment – Requirements and guidelines
▶ ISO 50001:2011, Energy Management – Requirements with guidance for use
▶ ISO 20121:2012, Event Sustainability Management Systems – Requirements with guidance for use

▶ PAS 2050:2011, Specification for the assessment of the life cycle greenhouse gas emissions of goods and services
▶ PAS 2060:2014, Specification for the demonstration of carbon neutrality.

Exam tip

You are not expected to remember any specific ISO/PAS document.

EC 1.3

Environmental impact assessments

Learning outcomes

3.1 Explain the reasons for carrying out environmental impact assessments ☐

3.2 Describe the types of environmental impact ☐

3.3 Identify the nature and key sources of environmental information ☐

3.4 Explain the principles and practice of impact assessment ☐

3.1 **Explain the reasons for carrying out environmental impact assessments**

> **Key revision points**
> ▶ Meaning of aspects, impacts (ref ISO 14001)
> ▶ Aims and objectives of impact assessment
> ▶ Cradle-to-grave concept (life cycle analysis)

Meaning of aspects, impacts

Aspect

> *[T]he element of an organisation's activities or products or services that can interact with the environment; significant aspects will have a significant environmental impact.*

Aspect can come in many forms and all will have some detrimental effect on the environment. For example:

▶ hazardous materials;
▶ emissions to land, air or water;
▶ waste whether it is solid, liquid or any other medium;
▶ nuisance emissions;
▶ raw material usage; and
▶ energy and land usage.

Impact

> *[A]ny change to the environment, whether adverse or beneficial, wholly or partially resulting from an organisation's environmental aspects.*

For example, impacts may include noise nuisance or health effects.

A further example of aspects and impacts can be seen below (3.2).

Circumstances that would require a company to review any previous assessments of its aspects:

▶ Significant changes to some of the elements such as changes in raw material source or change of product.

▶ Changes in plant or location, the availability of new processes and technologies or introducing recycling opportunities.

▶ Introduction of new legislation.

▶ Market and public pressures.

▶ Evidence of environmental damage or accidents.

▶ Results from audits.

▶ Annual reviews as part of an EMS.

▶ Industry or trade body recommendations.

▶ Pressure from non-governmental organisations (NGOs).

Aims and objectives of impact assessment

An environmental impact assessment (EIA) has been defined as:

A systematic process to identify, predict and evaluate the environmental effects of proposed actions and projects.

An EIA is a planning tool that identifies the consequences of any development prior to the start of the project.

[T]he process of identifying, predicting, evaluating and mitigating the biophysical, social and other relevant effects of development proposals prior to major decisions being taken and commitments made.

The EIA process derives from European law. The European law basis is **Directive 85/337**, the Assessment of the Effects of Certain Public and Private Projects on the Environment as amended by **EC Directive 97/11/EC**. The Directive is mainly implemented in UK legislation through the **Town and Country Planning (Environmental Impact Assessment) Regulations 2011**.

The International Association for Impact Assessment (IAIA) is the leading global network on best practice in the use of impact assessment for informed decision making regarding policies, programs, plans and projects.

The aims/objectives can be split into immediate and long term:

Table 3.1 Aims and objectives of EIA

Immediate aims/objectives of EIA are to:	Long term aims/objectives of EIA are to:
▶ Improve the environmental design of the proposal	▶ Avoid irreversible changes and serious damage to the environment
▶ Check the environmental acceptability of the proposals compared to the capacity of the site and the receiving environment	▶ Safeguard valuable resources, natural areas and ecosystem components
▶ Ensure that resources are used appropriately and efficiently	▶ Enhance the social aspects of proposals
▶ Identify appropriate measures for mitigating the potential impacts of the proposal	▶ Protect human health and safety
▶ Facilitate informed decision making, including setting the environmental terms and conditions for implementing the proposal	

Cradle-to-grave concept (life cycle analysis)

Life Cycle Assessment (LCA) (also known as life cycle analysis) is one of a number of environmental management techniques. It is a cradle-to-grave analysis tool that analyses all of the inputs, processes and outputs of a product/system over its entire life cycle. ISO 14040 defines LCA as:

Consecutive and interlinked stages of a product system, from raw material acquisition or generation from natural resources, to final disposal.

The preparation of an inventory analysis includes quantifying releases of greenhouse gases at each stage of an LCA.

Inputs – Manufacture – Outputs – Use – Disposal

By interpreting the inventory, the magnitude of total contributions of releases at each stage of the product life cycle can be understood.

LCA results are often categorised in several ways, including:

- global warming potential;
- acidification;
- outrophication (*excessive richness of nutrients in a lake or other body of water, frequently due to run-off from the land, which causes a dense growth of plant life*);
- tropospheric ozone creation; and
- stratospheric ozone depletion.

LCA is a specialist method for examining the costs and burdens that activities or products place on the environment.

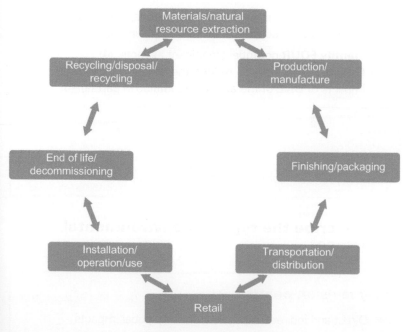

Figure 3.1 Stages of the Life Cycle Assessment

LCA is the *compilation and evaluation of any input, output and their potential environmental impacts of product systems throughout its life cycle.* There are four stages:

Stage 1 – Definition of goal and scope

Stage 2 – Inventory analysis

Stage 3 – Impact assessment

Stage 4 – Interpretation

Successful implementations of LCAs have been achieved in many industries, including automobile, airline and consumer goods companies. ISO 14040:2006 and 14044:2006 are the standards for LCA.

Revision exercise

A large high street chain store selling clothes, household goods and food wants to reduce its environmental impact:

(a) **Identify FOUR** potential significant aspects. (4)
(b) **Outline FOUR** ways to reduce the associated environmental impacts of **ONE** of the aspects identified in part (a). (4)

Exam tip

Review your own company's aspects and impacts.

3.2 Describe the types of environmental impact

Key revision points

▶ Direct and indirect impacts, including global impacts
▶ Contamination of atmosphere
▶ Contamination of land

▶ Contamination of aquatic environment
▶ Positive and negative effects on the community including visual impact and loss of amenity
▶ Positive and negative effects on the ecosystem

Direct and indirect impacts, including global impacts

Pollution pathways (routes by which pollutants are distributed through the environment, e.g., air condition ducts, food chains, waterways) need to be understood to be able to identify direct and indirect impacts, including global impacts.

▶ Recognise that the pathway is the link between a source and a receptor in spreading pollution.
▶ Pathways can be through air, land or water.
▶ Breaking the pathway will avoid pollution spreading.
▶ Pollutants can be changed in concentration or chemical form within a pathway and bioaccumulation may occur.
▶ Such mechanisms within the pathway determine the dose of pollutant delivered from source to receptor.

Impacts can be classed as **direct** and **indirect** impacts of the activities of an organisation.

For example, an organisation will use energy, i.e., electricity. The indirect generation of this, say from a coal-fired power plant,

Table 3.2 Direct and indirect impacts

Direct impacts	Indirect impacts
Activities over which a company can be expected to have an influence and control, for example, emissions from processes	Actual or potential activities over which the organisation can be expected to have an influence, but no control, for example, supply chain controlled aspects, customer controlled aspects, aspects managed elsewhere within the same company

will cause air pollution (carbon dioxide, sulphur dioxide, nitrogen oxides, etc.). The creation this impact has is an indirect result of the organisation but it is a direct result of the power plant.

Contamination of atmosphere

Table 3.3 Contamination of atmosphere

Aspects	Impacts
Air emissions	Air pollution
▶ Chemical (fumes, VOC, etc.)	▶ Ozone, etc.
▶ Particulate/dust	▶ Air quality
	▶ Human
	▶ Extreme weather events
	▶ Coastal erosion
	▶ Rising sea levels
	▶ Acid rain
	▶ Photochemical smog

Contamination of land

Table 3.4 Contamination of land

Aspects	Impacts
Releases to land/waste by-products	▶ Cost of disposal
	▶ Contamination of land
	▶ Effect on flora and fauna
	▶ Effect on ecosystem
	▶ Crops

Contamination of aquatic environment

Table 3.5 Contamination of aquatic environment

Aspects	Impacts
Releases to water/discharge	▶ Cost of treatment/filtration
	▶ Contamination of water courses
	▶ Effect on aquatic life/ecosystem

Positive and negative effects on the community including visual impact and loss of amenity

Table 3.6 Positive and negative effects on the community

Aspects	Impacts
Nuisance (statutory)	▶ Negative impact on communities
▶ Odour	▶ Light pollution
▶ Noise/vibration	▶ Effects on fauna and flora
▶ Visual impact	
▶ Lighting glare (nuisance)	
▶ Loss of amenity	

Whilst there are many negative impacts on the community, including those listed above, positive impacts can also occur, including, for local residents, the development of jobs, better road structures, building of public parks, etc.

Factors to consider when evaluating the significance of environmental impacts:

▶ size or magnitude of the impact (local vs global);
▶ nature (toxicity, for example);
▶ frequency of occurrence;
▶ likelihood and duration of impact;
▶ severity of consequences;
▶ sensitivity of the receiving environment;
▶ factors which were relevant to the receptor;
▶ the pathway between source and receptor may be significant as may be the extent to which the impact is reversible;
▶ legal requirements; and
▶ contractual obligations which may be relevant as can the importance of interested parties and the costs involved.

Aspects and impacts on the ecosystem

Table 3.7 Aspects and impacts on the ecosystem

Aspects	Impacts
Use of raw materials	▶ Cost of materials
	▶ Indirect impacts of supply of materials
	▶ Natural resource depletion
	▶ Indirect impact of manufacturing process
Use of energy	▶ Cost of supply
	▶ Indirect impacts of energy production
Use of water	▶ Cost of incoming supply
	▶ Indirect impacts of supply of water

Ecosystems are divided into terrestrial or land-based ecosystems and aquatic ecosystems in water, for example, forests, grasslands, deserts, wetlands or coastal areas. The nature of the ecosystem is based on its geographical features, for example, hills, mountains, plains, rivers, lakes, coastal areas or islands.

The living part of the ecosystem is referred to as its biotic component. All the living organisms in an area live in communities of plants and animals. They interact with their non-living environment, and with each other at different points in time for a large number of reasons. These features create conditions that support a community of plants and animals that evolution has produced to live in these specific conditions.

Ecosystems may be adversely affected by the activities of the organisation. Damage to a number of linked habitats will eventually have a damaging impact on the surrounding ecosystems.

Examples include if a river becomes polluted, then this could cause harm to the fish and other wildlife, damage crops and leak into the food chain.

Deforestation will cause:

- damage to local habitats and wildlife;
- impacts on the local population;
- impacts on lifestyle and food sources;
- economic impacts that can affect tourism and jobs;
- reduced carbon storage and oxygen production;
- impacts on the local climate;
- reduced recycling of nutrients;
- loss of biodiversity;
- loss of rare tree species;
- disruption to the regulation of the water cycle;
- increased soil erosion; and
- desertification.

Revision exercise

(a) **Outline** the difference between direct and indirect impacts. (2)
(b) **Identify** impacts that cause:
  (i) Atmospheric pollution. (2)
  (ii) Aquatic contamination. (2)
  (iii) Land contamination. (2)

Exam tip

Identify the impacts and aspects associated with your place of work.

3.3 Identify the nature and key sources of environmental information

---

Key revision points

- Internal to the organisation
- External to the organisation

Key sources internal to the organisation include:

▶ audit and investigation reports;
▶ maintenance records;
▶ inspections;
▶ job/task analysis;
▶ incident data;
▶ use of environmental monitoring data to evaluate risk; and
▶ raw material usage and supply.

Key sources external to the organisation include:

▶ manufacturer's data;
▶ legislation (national and international);
▶ government or regulatory bodies;
▶ environment agencies;
▶ trade associations;
▶ international, European and British standards;
▶ internet; and
▶ encyclopedias.

Revision exercise

Identify FOUR sources of information:

(a) internal (4)
(b) external (4)

of an organisation that could be consulted when undertaking environmental impact assessments.

Exam tip

Review what current sources of information your organisation is using for environmental management – are there any additional sources that might be useful?

3.4 Explain the principles and practice of impact assessment

> ### Key revision points
>
> ▶ Linked to the initial environmental review
> ▶ Consider normal and abnormal conditions, incidents, accidents and potential emergency situations; past activities, current activities and planned activities
> ▶ Concept of source, pathway and receptor when assessing environmental risk
> ▶ International impacts; resource abstraction; pollution from mining, transport and processing; waste disposal
> ▶ Identifying receptor at risk
> ▶ Identification of aspects/impacts
> ▶ Evaluating impact and adequacy of current controls
> ▶ Supplier selection, transport issues
> ▶ Recording significant aspects format
> ▶ Reviewing: reasons for review

Environmental Impact Assessment (EIA)

Whether undertaking an Environmental Impact Assessment (EIA) or simply an Environmental Aspects and Impacts Register it is important to be able to identify all aspects and impacts. During the EIA the process normally followed will aid the identification of the impacts. The EIA process follows a number of commonly accepted steps:

▶ factors to consider for an impact assessment;
▶ proposal identification;
▶ screening;
▶ scoping;
▶ impact analysis;
▶ mitigation;
▶ avoidance;

▶ environmental statement;
▶ decision; and
▶ links to the initial environmental review.

Factors to consider for an impact assessment:

▶ screening and scoping stages;
▶ responses and issues raised by statutory consultees;
▶ findings of baseline surveys;
▶ characterisation of the development during construction, operation and decommissioning;
▶ location of processes and possible alternatives;
▶ materials (storage, use of resources, etc.);
▶ topography, geology and hydrogeology;
▶ socio-economic factors (including neighbours, demographics, etc.);
▶ sites of specific scientific interest, wildlife, etc.;
▶ emission and mitigation techniques in relation to air, water and land pollution;
▶ land use, archaeology, visual aspects;
▶ emergency procedures;
▶ security;
▶ transport effects;
▶ energy emissions (noise, vibration, heat, light, radiation);
▶ drainage and surface water pollution; and
▶ preparation of technical and non-technical reports.

Proposal identification

At this initial stage a large number of decisions are made at the project identification and proposal development stage, for example impacts, resource abstraction, pollution from mining, transport and processing and waste disposal. By identifying the receptors at risk, for example flora, fauna, watercourse, local residents including indigenous peoples, etc., it may be possible to reduce the impacts and in some cases remove them altogether.

Screening

Screening is undertaken to determine if a development proposal requires an EIA. Local legislation will need to be followed (for example: 'Council Directive 97/11/EC amending Directive 85/337/EEC'. The criteria may include:

- ▶ development characteristics;
- ▶ characteristics of the location; and
- ▶ characteristics of potential effects.

Scoping

During the scoping stage, the key issues to be addressed are identified. A scoping process will normally involve:

- ▶ **Consultation** with all relevant stakeholders.
- ▶ **Analysis** of the issues identified to determine which are likely to be significant and therefore must be included within the scope of the EIA.
- ▶ **Negotiation** with the decision makers and other interested parties to refine the scope of the EIA.
- ▶ **Feedback** of any positive or negative findings to all interested parties.

There are a number of issues that are considered (to varying depths) within the scope of the majority of EIAs. These include:

- ▶ air and water quality;
- ▶ land use and ground conditions;
- ▶ landscape and visual appearance;
- ▶ ecosystem;
- ▶ traffic and transport, including impact of noise; and
- ▶ archaeology and cultural heritage.

Impact analysis

This stage will involve identifying and evaluating the likely impacts of the development proposal. This is likely to involve:

- ▶ consideration of scale and severity of the impact;
- ▶ probability of occurrence;

▶ duration of impact (or business concerns);
▶ sensitivity of receiving environment;
▶ consideration of legal or contractual requirements;
▶ concern of interested parties and any effect on public image;
▶ activities of suppliers, degrees of influence on product design; and
▶ supplier selection, transport issues – distribution and staff movements.

Mitigation

Mitigation is trying to compensate for damage, e.g., by creating a new wetland.

Avoidance

Avoidance is finding another way or putting controls in place.

Environmental Statement

The environmental statement is a legal document (for example identified in article 5 (3) of the EIA Directive) and must include, at least, the following information:

▶ A description of the project comprising information on the site, design and size of the project.
▶ A description of the measures envisaged avoiding, reducing and, if possible, remedying significant adverse impacts.
▶ The data required to identify and assess the main effects that the project is likely to have on the environment.
▶ An outline of the main alternatives studied by the developer and an indication of the main reasons for this choice, taking into account the environmental effects.
▶ A non-technical summary of the information mentioned in previous bullet points.

Decision

This point in the process is where the development is either granted planning permission or not and is likely to involve local government officials.

Linked to the initial environmental review

An initial environmental review would include reviewing/identifying:

▶ breaches of legislation;
▶ impact assessments;
▶ effluent treatment and discharge;
▶ emissions to the atmosphere;
▶ natural habitat;
▶ noise nuisance;
▶ production and operations;
▶ resources; and
▶ materials – storage, use and waste.

Consider normal and abnormal conditions, incidents, accidents and potential emergency situations; past activities, current activities and planned activities

The organisation's aspects (activities or products or services that can interact with the environment) can lead to different types of environmental impacts (any change to the environment, whether adverse or beneficial). The activities or products or services can be considered as normal or abnormal operations or emergency conditions.

Table 3.8 Aspect and impacts

Activity type	Aspect	Impacts
Normal operation	Using heating oil for factory	Air pollution and potential climate change
Abnormal operation	Filling the oil tank	Pollution of a nearby watercourse or pollution into site drains if oil/fuel spills
Emergency condition	Leaks from bunded oil container or fire	In the event of a vehicle fire, this could result in release of toxic fumes

65

It is necessary for an organisation to consider the environmental consequence of each situation. This will involve some form of the qualitative assessment procedure.

Concept of source, pathway and receptor when assessing environmental risk

International impacts; resource abstraction; pollution from mining, transport and processing; waste disposal

Without the presence of all three elements of the source–pathway–receptor model, pollution cannot occur. The source is the process/activity where the aspect started, the pathway is the transmission route of the aspect and the receptor is where the process/activity will have the effect.

Table 3.9 Examples of source–pathway–receptor

Sources
- ▶ Exhaust/chimney emissions
- ▶ Welding fumes
- ▶ Machinery (creating noise and vibration)
- ▶ Chemical spill into river or on grassland
- ▶ Fertiliser on a field running off and polluting a watercourse – eutrophication

Pathways
- ▶ Air
- ▶ Land
- ▶ Water
- ▶ Biological transfer through the food chain

Receptors
- ▶ Atmosphere
- ▶ Land
- ▶ Watercourse (including rivers and groundwater)
- ▶ Building
- ▶ Human beings/local residents
- ▶ Ecosystem (including flora, fauna)

Table 3.10 Example for car's source–pathway–receptor

Source	Pathway	Receptor
Exhaust fumes from a car	Airborne – inhalation of fumes	Human – health effects

Table 3.11 Example for petrol station's source–pathway–receptor

Source	Pathway	Receptor
Underground fuel tank	Base flow and discharge through ground water flow	Local water courses/rivers
Fuel dispenser	Volatilisation of spilt hydrocarbons – air contamination – inhalation of vapours	Humans/food chain
Spills from customer/suppliers activities	Forecourt drains	Local water courses/rivers
Exhaust fumes from vehicles	Airborne – inhalation of fumes	Humans/food chain

For example, for a petrol station:

▶ The underground tank may leak – this will depend on the age of the tank or if it is bunded, etc.
▶ Fuel dispenser leakage will lead to volatilisation (i.e., fuel releases vapour) of spilt hydrocarbons leading to air contamination leading to inhalation of vapours.
▶ Spillage from customer/suppliers activities could result in this entering the watercourses. Alternatively, this could result in a fire or explosion.
▶ Exhaust fumes from vehicles using the forecourt – fuel fumes will become airborne and may have an impact on human health and/or climate change.

Identifying receptors at risk

These include:

▶ flora;
▶ fauna;

▶ watercourse (ground water and surface water);
▶ housing or buildings; and
▶ local residents including indigenous peoples, etc.

The following factors should be considered when evaluating the impact and adequacy of current controls:

▶ consideration of scale and severity of the impact;
▶ probability of occurrence;
▶ duration of impact (or business concerns);
▶ sensitivity of receiving environment;
▶ consideration of legal or contractual requirements;
▶ concern of interested parties; and
▶ effect on public image.

Identification of aspects/impacts

The identification of aspects/impacts may include:

▶ atmospheric pollution;
▶ aquatic contamination;
▶ land contamination;
▶ resource depletion;
▶ effects on the community;
▶ effects on the ecosystem; and
▶ loss of raw materials and natural resources.

Fire will have a detrimental environmental impact, including damage to flora and fauna, property damage, the release of harmful toxins within the smoke.

Evaluating impact and adequacy of current controls

It is necessary for an organisation to consider the environmental consequence of each situation. This will involve some form of quantitative assessment procedure. For example, in a health and safety risk assessment likelihood × consequence will provide a risk rating. The following tables show how this procedure develops.

Table 3.12 Risk matrix 1

Likelihood of occurrence, the aspect is:	Score
Very likely – 1 in 100	5
Likely – 1 in 1,000	4
Fairly likely – 1 in 10,000	3
Unlikely – 1 in 100,000	2
Very unlikely – 1 in a million	1
Consequence, will the aspect be:	**Score**
Major and have a serious effect on the environment	5
Significant and have a noticeable effect on the environment	4
Moderate with a known effect on the environment	3
Limited with little effect on the environment	2
Minimal with very little effect on the environment	1

Table 3.13 Risk matrix 2

Score (Likelihood × consequence)	Timeframe	Action
20–25	Immediate	Activities should not continue until the risk has been reduced
10–16	Short term	Measures should be taken within one month
5–9	Medium term	Measures should be taken within the next six months
3–4	Long term	Measures should be taken within the next 12 months
1–2	None	Ongoing active monitoring as part of day-to-day operations

Note: This is only one of many ways to quantify the level of environmental risk.

Supplier selection

Supplier selection is the process by which organisations will evaluate current and future suppliers. The organisation may adopt an environmental procurement policy, for example, only using suppliers that can demonstrate their environmental management commitments.

Recording significant aspects format

An Environmental Aspects and Impacts Register is an important document that should normally identify all significant aspects and impacts.

Formats vary from organisation to organisation (as do health and safety risk assessments) within an Environmental Aspects and Impacts Register regarding the layout and detail included.

Consideration must be made for the following:

▶ It should show the activities for each of the related aspects and impacts, plus suitable current, and where relevant, future controls.

▶ When assessing the aspects and impacts, the source–pathway–receptor principle should be considered.

In addition, it may include the list in Table 3.14:

Table 3.14 Recording aspects

Activity	What is the process, task or project to be undertaken?
Conditions	Will it address Normal, Abnormal and Emergency conditions?
Aspects	How will the activities, products or services interact with the environment?
Impacts	What changes will there be to the environment, whether adverse or beneficial?
	It will be necessary to indicate if the impact is a direct or indirect cause of the activity/aspect.
Risk matrix	Will it be quantitative (numerical) or qualitative (high, medium, low)? It may also be worth considering including other aspects to the risk matrix; for example legal compliance.
Controls	What controls will be put in place to reduce or minimise the impacts?
Priority	A numerical system may be used to:
	Identify where further controls are needed
	Who will be responsible?
	When will the additional controls be carried out?

Environmental Aspect and Impact Register – Simple Example

Table 3.15 EIA part 1

Activity	Activity type N/A/E	Aspect	Impacts	Control/mitigation	Risk matrix	Priority
Office heating – use	Normal operation	Using heating oil for factory	Air pollution and potential climate change	▲ Record usage per month ▲ Signage used to remind employees to turn off lights, etc. ▲ Purchasing policy includes purchase of low energy equipment	Low	None
Office heating – filling tank	Abnormal operation	Filling the oil tank	Pollution of a nearby watercourse or pollution into site drains if oil/fuel spills	▲ Follow safe system of work ▲ Ensure only authorised persons to fill tank	Med	1
Office heating – emergency	Emergency condition	Leaks from bunded oil container or fire	In the event of a vehicle fire, this could result in the release of toxic fumes	▲ Regular checks of the tank, pipes and connectors	High	2

Table 3.16 EIA part 2

Activity	Priority	Legislation	Further controls/mitigation	Person responsible	Date
Office heating – use	none	n/a	n/a	n/a	n/a
Office heating – filling tank	1	Local Regulations/ Acts	Supervise/monitor filling of tank and keep records of who filled, amount, who supervised	A N Other, office manager	June 20XX
Office heating – emergency	2	Local Regulations/ Acts	Develop a system to record checking of the tank. Have available absorbent mats, etc. in the event of leaks	A N Other, office manager	July 20XX

Reviewing: reasons for review

As with health and safety risk assessments, there are various reasons why an organisation may be required to review its aspects and impacts. These include:

▶ any environmental incidents – for example, spillage of heating oil;
▶ purchase of or change to any equipment;
▶ change to any processes;
▶ changes to key staff;
▶ legislative changes or new international conventions, etc.;
▶ after a given period of time, for example, one year; and
▶ prosecutions or legal notices.

Revision exercise

A new cement works is being planned for construction.

(a) **Outline** the main aspects associated with the proposal. (4)
(b) **Outline** the main potential impacts associated with the proposal. (4)

Exam tip

The above question could have selected any new works – would the answer have changed for a new wind farm?

Control of emissions to air

Learning outcomes	
4.1 Outline the principles of air quality standards	☐
4.2 Outline the main types of emissions to atmosphere and the associated hazards	☐
4.3 Outline control measures that are available to reduce emissions	☐

4.1 Outline the principles of air quality standards

> **Key revision points**
>
> ▶ Meaning, uses of and the relationship between ppm and mgm-3 (**mg.m³**)
> ▶ The potential effects of poor air quality
> ▶ The role of air quality standards and controls on quality and impurities

Meaning, uses of and the relationship between ppm and mg.m³

Substances that exist as a gas or vapour at normal room temperature and pressure can be expressed in parts per million (ppm):

▶ **ppm** = air pollutant concentration, in parts per million by volume

Compounds that do not form vapours at room temperature and pressure are expressed in mg.m³ or µg.m³.

Parts per million as a percentage

1,000,000 ppm	100 per cent
100,000 ppm	10 per cent
10,000 ppm	1 per cent

They are measured at: 'per cubic metre of air at sea level atmospheric pressure'.

▶ **mg.m³** = milligrams of pollutant
▶ **µg.m³** = micrograms of pollutant
▶ **ng.m³** = nanograms of pollutant

Common units

m	one thousandth (10^{-3})
µ	one millionth (10^{-6})
n	one billionth (10^{-9})

Particles consist of solid matter of all sizes from less than 0.001 micron to greater than 100 microns.

▶ **PM 2.5** = particle size less than 2.5 μm
▶ **PM 10** = particle size between 2.5 and 10 μm

The potential effects of poor air quality

The effects of poor air quality are wide ranging, including:

▶ Acid rain can be a trans-boundary pollutant.
▶ Acid rain damages water and plant life.
▶ Airborne pollutants can also deposit in water and affect the organisms living there.
▶ Airborne pollutants can build up in concentration and affect the health of animals that feed on the plants.
▶ Airborne pollutants can cause acid rain, which can damage man-made structures by eroding stonework and also eroding steel structures.
▶ Airborne pollutants can cause photochemical smog.
▶ Emissions of greenhouse gases cause global warming.
▶ There are human health effects such as asthma.

The role of air quality standards and controls on quality and impurities

The UK government, like those of other countries, has regulated air quality objectives covering various air pollutants, against which local authorities are required to measure local air quality.

Revision exercise

Two standard units used to measure air quality are ppm and mg.m^3.

(a) **Give** the meaning of EACH unit. (4)
(b) **Outline** the circumstances in which each may be used AND give an example in EACH case. (4)

Exam tip

Research your place of work to discover if they are releasing any common air pollutants.

Study point

The following table identifies common air pollutants, describes their health effects and explains where they can be found.

You are not expected to know all of this detail, but to be able to give an appropriate example of common air pollutant(s).

Table 4.1 Common air pollutants

Common air pollutants	Health effects	Source
Regulated pollutants in UK		
SMOG (SMoke & fOG)	Dense **SMOG** covered Greater London between 5 and 8 December 1952, accompanied by a sudden rise in mortality that far exceeded anything previously recorded during similar periods of smog The Ministry of Health's committee later estimated that between 3,500 and 4,000 more people had died	The combination of smoke and natural fog It has been applied to the brown haze created by emissions of nitrogen oxides and hydrocarbons from motor vehicles in strong sunlight in cities such as Los Angeles. This is more accurately referred to as photochemical smog, i.e., chemical reaction of air pollutants under certain climatic conditions
Nitrogen dioxide (NO_2)	**Nitrogen dioxide** can irritate the lungs and lower resistance to respiratory infections such as influenza Continued or frequent exposure to concentrations much higher than those	Nitric oxides (NO) are mainly derived from road transport emissions and other combustion processes such as the electricity supply industry. Nitric oxide is not considered to be harmful to health

Common air pollutants	Health effects	Source
	normally found in the ambient air may cause increased incidence of acute respiratory illness in children	However, once released to the atmosphere, NO is usually very rapidly oxidised, mainly by ozono (O_3), to nitrogen dioxide (NO_2), which can be harmful to health
		NO_2 and NO are both oxides of nitrogen and together are referred to as nitrogen oxides (NOx)
Particulates (PM_{10} & $PM_{2.5}$)	**Particles** are measured in a number of size fractions according to their mean aerodynamic diameter	Road traffic is a major source of particulate PM_{10}. Combustion sources (such as road traffic); secondary particles, mainly sulphate and nitrate formed by chemical reactions in the atmosphere, and often transported from far across Europe; coarse particles, suspended soils and dust (e.g., from the Sahara), sea salt, biological particles and particles from construction work
	Fine particles can be carried deep into the lungs where they can cause inflammation and a worsening of the condition of people with heart and lung diseases	
Sulphur dioxide (SO_2) A regulated pollutant in UK.	**Sulphur dioxide**, even moderate concentrations, may result in failing lung functions of those suffering from asthma	Industrial emissions are the main source of sulphur dioxide pollution at ground level, normally from plume grounding as a result of adverse weather conditions and/or stack design factors
	Tightness in the chest and coughing occur at high levels and lung function of asthmatics may be impaired to the extent that medical help is required	Globally, much of the SO_2 in the atmosphere comes from natural sources. A common man-made source is power stations burning fossil fuels, principally coal and heavy oils
	Sulphur dioxide pollution is considered more harmful when particulate and other pollution concentrations are high	Widespread domestic use of coal can also lead to high local concentrations of SO_2

(Continued)

Table 4.1 (Continued)

Common air pollutants	Health effects	Source
Carbon monoxide (CO)	**Carbon monoxide** gas prevents the normal transport of oxygen by the blood This can lead to a significant reduction in the supply of oxygen to the heart, particularly in people suffering from heart disease	Road traffic is a major source of carbon monoxide pollution. It is produced by incomplete, or inefficient, combustion of fuel It is predominantly produced by road transport, in particular, petrol-engine vehicles Industrial emissions can be significant, but are normally emitted through tall stacks, which give dispersion away from ground level
Benzene (C_6H_6)	**Benzene** has possible chronic health effects including cancer, central nervous system disorders, liver and kidney damage, reproductive disorders and birth defects	Benzene is a volatile organic compound (VOC) which is a minor constituent of petrol Main sources of benzene in the atmosphere are from industry and traffic emissions
1,3-butadiene (C_4H_6)	**1,3-butadiene** has possible chronic health effects including cancer, central nervous system disorders, liver and kidney damage, reproductive disorders and birth defects	1,3-butadiene is a VOC emitted into the atmosphere principally from fuel combustion of petrol and diesel vehicles 1,3-butadiene is also an important chemical in certain industrial processes, particularly the manufacture of synthetic rubber
Lead (Pb)	**Lead**, even small amounts, can be harmful, especially to infants and young children. In addition, lead taken in by the mother can interfere with the health of the unborn child	A major source of lead at ground level was from petrol-engine vehicle exhausts, but as a result of the introduction of lead-free petrol, this source is no longer significant

(Continued)

Common air pollutants	Health effects	Source
	Exposure has also been linked to impaired mental function, visual-motor performance and neurological damage in children as well as impaired memory and attention span	Lead is main metal pollutant in the air In recent years industry, in particular, secondary non-ferrous metal smelters, has become the most significant contributor to emissions of lead

Unregulated pollutants in UK

Common air pollutants	Health effects	Source
Ozone (O_3)	**Ozone** is associated with slight irritation to the eyes or nose Very high levels of exposure (in excess of 10 times the proposed objective level) over several hours can cause damage to the airway lining followed by inflammatory reaction	Ozone is a secondary air pollutant. It is not emitted by any process but is formed as a result of complex chemical reactions on other air pollutants, particularly in the presence of strong sunlight Source pollutants, such as nitrogen dioxide and hydrocarbons, are emitted from traffic and industry It is recognised, therefore, that local or even national action may not be sufficient to reduce ozone levels; the source pollutants can originate a considerable distance away
Polycyclic Aromatic Hydrocarbons (PAHs)	**PAH** exposure is associated with an increased incidence of tumours of the lung, skin and other sites, with lung cancer most obviously linked to exposure through inhaled air Objective level of 0.25 ng/m³ (nanograms per metre cubed) as an annual average is considered	Main sources are domestic coal and wood burning fires and industrial processes such as coke production Road transport is the largest source for total PAHs

(Continued)

Table 4.1 (Continued)

Common air pollutants	Health effects	Source
	to represent a risk to health so small as to be undetectable	
Cadmium (Cd)	**Cadmium** is bio-persistent and derives its toxicity from its chemical similarity to zinc, which is an essential micronutrient Long-term exposure can cause renal misfunction High levels are associated with lung disorders and bone defects	Cadmium is produced as an inevitable by-product of zinc, and sometimes lead refining It is mainly used in high-performance nickel/cadmium batteries but is also a good corrosion resistance coating. Other uses are as pigments, stabilisers for PVC, in alloys and electronic compounds Cigarette smoking and some foods can be a source
Arsenic (As)	**Arsenic** is highly toxic in its organic form It may be beneficial in small doses, but is generally considered to be carcinogenic to the lungs and skin	Natural sources such as volcanic eruptions and forest fires Man-made emissions are likely to arise from coal burning, industrial waste disposal and the application of agricultural chemicals containing arsenic, and the burning of wood with arsenic-containing preservatives Cigarette smoking can be a significant source; food and water are other sources
Nickel (Ni)	**Nickel** compounds generally exhibit a low acute toxicity Nickel and its water-soluble salts are potent skin sensitisers and are restricted for jewellery use	Used in the production of stainless steels Nickel alloys and plating are commonly found in vehicles, tools, electrical and household goods, jewellery and coinage

Common air pollutants	Health effects	Source
	where there may be in direct contact with the skin	From combustion of coal and oil for heat and power generation and the incineration of wastes and sewage sludge
		Cigarette smoking can be a significant source; food is another source
Mercury (Hg)	**Mercury** is a toxic substance with no known function in human biochemistry or physiology	It occurs naturally in the atmosphere from degassing of the earth's crust, emissions from volcanoes and evaporation from natural bodies of water
	Inorganic poisoning can cause tremors and spontaneous abortion	Industrial processes and in products such as batteries, lamps and thermometers
	Mercury methyl compounds cause damage to the brain and central nervous system	It is associated with chlorine manufacture
		Used in dentistry as an amalgam for fillings and by the pharmaceutical industry

4.2 Outline the main types of emissions to atmosphere and the associated hazards

Key revision points

▶ Types of emissions

Common types of emissions

Table 4.2 Common types of emissions

Type of emission	Description
Gaseous	Substances which remain in the gaseous phase at normal atmospheric temperatures and pressures, e.g., carbon dioxide, nitrogen, ozone. For example: from the combustion of oil, coal, gas, etc.
Vapour	The gaseous state of materials which are liquid at normal temperature and pressure. For example water vapour is the most important greenhouse gas. This is part of the difficulty with the public and the media in understanding that 95 per cent of greenhouse gases are water vapour
Odours	These are caused by one or more volatilised chemical compounds, often at low concentrations
Mists	Are formed when vapours condense and are composed of fine liquid droplets in the range 0.01 to 10 microns
Fumes	Fumes are small solid particles produced by condensation of vapours or gaseous combustion products (i.e., cooling of combustion products from hot processes). Particle size is in the range 0.1 to 1 micron
Smoke	Airborne gas, solid and liquid particulates emitted when a material undergoes combustion or pyrolysis. Particles in the range 0.1 microns to 10 microns are seen as smoke
Dust	Consists of any size or shape of particle. Particle sizes capable of inhalation are up to 10 microns; particle sizes of less than 7 microns are capable of penetrating lung tissue
Grit	Grit are particles exceeding 76 microns in diameter
Fibres	Solid particles with an increased aspect ratio (the ratio of length to width); for example, asbestos
Fugitive emissions	Emissions of gases or vapours from pressurised equipment due to leaks and other unintended or irregular releases of gases, mostly from industrial activities

Common pollutants

Table 4.3 Common types of pollutants

Common pollutants	Example
Sulphur compounds	The two major sulphur oxides (SO_x) are sulphur dioxide (SO_2) and sulphur trioxide (SO_3). When sulphur dioxide combines with oxygen (O_2) in the air some sulphur trioxide is slowly formed
Nitrogen compounds	Nitrogen oxides (NO_x) consist of nitric oxide (NO), nitrogen dioxide (NO_2) and nitrous oxide (N_2O). They are formed when nitrogen (N_2) combines with oxygen (O_2)
Halogens and their compounds	Halogens is the collective term used for the elements bromine (Br), chlorine (Cl), fluorine (F) and iodine (I). A common compound includes chlorofluorocarbon (CFC), which is an organic compound that contains carbon, chlorine and fluorine
Metals and their compounds	Metals and their compounds include: ▶ Lead (Pb); ▶ Cadmium (Cd); ▶ Nickel (Ni); and ▶ Mercury (Hg)
Volatile organic compounds	Volatile organic compounds is a term used to describe a range of unrelated solvents and similar products that are liquids at normal temperature and pressure. They include: ▶ Acetone; ▶ Benzene; ▶ Chloroform; and ▶ Trichloroethane

Revision exercise

Identify the common types of emissions to the atmosphere. (8)

> **Exam tip**
>
> Having a practical example of the types of emission/common pollutants will help to remember these common pollutants.

4.3 Outline control measures that are available to reduce emissions

> **Key revision points**
>
> ▶ Control hierarchy
> ▶ Examples of technology

Control hierarchy

The hierarchy of control is a structured way to control hazards/ potential harm. The order is important – with the best option considered first, i.e., to eliminate the hazard/potential cause of harm. The principle of the hierarchy of control is to address the **source** of the emissions into the atmosphere first. If this cannot be achieved, or does not resolve the problem, then an attempt should be made to control the release of harmful substances along the transmission **path**. If this is effective, then the **receiver** is not affected.

Stages for an environmental hierarchy of control outlined below.

▶ **Eliminate/prevent pollution** by ceasing or changing the process or the materials used. For example:
  ▷ Instead of traveling to meetings, use telecommunication software.
▶ **Substitution** by replacing one substance with another that is less damaging in its effects or using substances less prone to causing emissions. For example:
  ▷ Instead of using organic solvents, use water-based detergents.

▶ **Minimise/reduce emissions** to the lowest practicable level by means of control devices, process management or choice of materials. For example:
  ▷ Lower process temperature to reduce vapour emissions.
▶ **Render any remaining emissions/discharge harmless** by reducing the polluting content. For example:
  ▷ Use scrubber systems.

Source, pathway, receptor

Table 4.4   Source–pathway–receptor

Source	Pathway	Receptor/receiver
Release of harmful emissions from vehicle/ power station	Air	▶ River ▶ Stream ▶ Lake ▶ Groundwater ▶ Humans ▶ Animals

Examples of technology

Different types of technology may be the Best Available Technique (BAT).

Table 4.5   Examples of technology

Examples of technology	Description
Bag filter	▶ Bag filter is used when the particles to be removed are larger than the pore size in the filter medium ▶ This traps the particles on the filter and cleaned air passes through, but as solids build up the flow rate is reduced ▶ At some stage the filter has to be cleaned by agitation ▷ reverse air pulse ▷ replacement

(Continued)

Table 4.5 (Continued)

Examples of technology	Description
Local exhaust ventilation with filtration	▶ The LEV will extract the pollutants and, prior to discharging them to the atmosphere, a suitable filter will be used to prevent the pollutants entering the atmosphere
Gravity separators	▶ Contaminated air is passed into a large chamber where the airflow speed is reduced, allowing particles and droplets to fall into a hopper under the force of gravity
Electrostatic precipitator	▶ Electrostatic precipitators rely on particles in an air stream gaining a charge as they pass between wires carrying a high voltage
	▶ Charged particles are then attracted to oppositely charged plates where they collect allowing the cleaned gas to discharge
	▶ Plates are cleaned by shaking or vibrating them with the particles falling into a trap
	▶ This equipment is used for particles such as dust, grit or fumes but not for particles that can form an explosive mixture in air
Wet scrubber: ▶ Acidified ▶ Basified	▶ A wet scrubber relies on the contaminated air flowing upwards against a cascade of water or an acid or alkaline solution
	▶ Soluble gases and particles are removed and the cleaned air flows out of the top
	▶ Horizontal plates or a fine spray give a large surface area for the water and this improves efficiency
	▶ Contaminated solution has to be removed from the base and treated separately
Absorption/ absorbers	▶ Chemical absorption is the most suitable method for the separation of CO_2 from exhaust gases
Water wall	▶ Water wall dust extractors are used to prevent dust entering the workplace and the atmosphere. The dust particles collect on the water wall

Greenhouse gases

Greenhouse gases emitted by human activities include:

▶ carbon dioxide (CO_2);
▶ methane (CH_4);

▶ nitrous oxide (N_2O);
▶ ozone (O_3); and
▶ fluorinated gases:
  ▷ hydrofluorocarbons (HFCs);
  ▷ perfluorocarbons (PFCs); and
  ▷ sulphur hexafluoride (SF_6).

Water vapour (H_2O) from the evaporation of water and/or sublimation of ice accounts for approximately 30–70 per cent of the greenhouse gases in the atmosphere.

Manufacturing activities' impact on the greenhouse effect

Table 4.6   Manufacturing activities' impact on the greenhouse effect

Direct	Indirect
▶ The release of greenhouse gases to the atmosphere including carbon dioxide, nitrous oxide, methane, water vapour, f-gases, etc.	▶ Use of electrical energy generated from fossil fuels at powers stations
▶ Process emissions/emission from on-site energy generation from fossil fuels	▶ Methane emissions from degradation of biodegradable wastes
▶ Emission from transportation systems, including both haulage and employee work-related transport	▶ Emissions caused by others in making products used in manufacturing
▶ Service emissions including leakages from refrigeration/cooling systems	▶ Emissions from cement manufacture for buildings, etc.
	▶ Removal of environmental sinks: forests, peat and soil

Greenhouse gases trap longwave thermal radiation, causing a warming effect upon the atmosphere. As concentrations increase, this warming effect increases to the point where there is an imbalance between incoming shortwave radiation and outgoing longwave radiation leading to net global warming of the atmosphere. The possible consequences of this warming include climate change, possible sea level rises, effects on agricultural and natural ecosystems, etc.

Nitrogen cycle

Nitrogen is released from both natural and man-made activities.

Step 1 – **Nitrogen fixation** – bacteria convert the nitrogen gas (N_2) to ammonia (NH_3) which the plants can use.

Step 2 – **Nitrification** – the ammonia converts into nitrite ions which plants take in as nutrients.

Step 3 – **Ammonification** – process in which bacteria breaks down organic matter into ammonia.

Step 4 – **Denitrification** – bacteria convert the nitrogen compounds back into nitrogen gas (N_2); this is then released back into the atmosphere to begin the cycle again.

Man-made activities:

Include:

▶ making and using nitrogen-based fertilisers and runoff into water;
▶ high-temperature processes causing oxidation of nitrogen gas to nitrogen oxide/nitrogen dioxide;
▶ discharge of sewage effluent containing nitrogen compounds into rivers;
▶ deforestation;
▶ industry; and
▶ farming methods including product farmed.

Difficulties in funding and maintaining equipment in some countries or environments

Equipment such as filtration units, local exhaust ventilation systems, gravity separators, electrostatic/magnetic precipitators, wet scrubbers, absorbers and water wall dust extractors can be expensive to purchase or fund initially. Further costs would include the installation and maintenance of the equipment. General day to day running costs of this equipment will be higher in harsh environments, for example very hot regions. This may become too expensive in some developing countries. If the correct equipment is not used and maintained correctly, this could cause harm to the workers and the local environment, through air pollution.

Revision exercise

(a) **Identify FOUR** atmospheric pollutants that may be released from a coal-fired power station. (4)

(b) **Outline** the control hierarchy for reducing air pollution emissions. (4)

Exam tip

If you are not familiar with the following types of technology, look up where they might be used:

▶ Filtration systems
▶ Separation technology
▶ Wet scrubbers
▶ Absorption
▶ Water walls

EC 1.5

Control of contamination of water sources

Learning outcomes	
5.1 Outline the importance of the quality of water for life	☐
5.2 Outline the main sources of water pollution	☐
5.3 Outline the main control measures that are available to reduce contamination of water sources	☐

5.1 Outline the importance of the quality of water for life

> **Key revision points**
>
> ▶ What is meant by safe drinking water
> ▶ The water cycle; sources of water such as groundwater, surface water and desalination
> ▶ Water for agriculture and industry
> ▶ Impact of water pollution on wildlife
> ▶ Over-abstraction; water conservation; balancing the different needs
> ▶ The potential effects of pollution on water quality

What is meant by safe drinking water?

Drinking water or potable water is water safe enough to be consumed by humans or used with low risk of immediate or long-term harm.

Safe drinking water must meet international or national standards for drinking water if it is to be used without the risk of immediate or long-term harm.

Drinking water is collected from a variety of sources:

▶ surface reservoirs;
▶ rivers; and
▶ groundwater – from underground strata (aquifers).

The recommendations from the World Health Organization's (WHO) Guidelines for Drinking-water Quality Report address the management of risk from hazards that may compromise the safety of drinking-water. The report describes the infectious water-borne diseases caused by pathogenic (harmful) bacteria, viruses and parasites. For example:

▶ Bacteria: E. coli, Legionella, Leptospira, Salmonella, Vibrio cholera, Typhoid, Para-Typhoid, etc.

▶ Viruses: Hepatitis A, Hepatitis E, Noroviruses, etc.
▶ Parasites: Protozoa (microscopic, one-celled organisms),
Helminthes (worm-like organisms), etc.

Study point

The following table identifies three types of characteristics that will have an impact on water. You are not expected to know all this detail but to be able to give appropriate examples of these characteristics.

Three types of characteristics (chemical, biological and physical) that will have an impact on water are:

Table 5.1 Characteristics that will have an impact on water

Type	Factor	Description
Chemical characteristics	▶ Salinity	Containing or impregnated with various types of salts
	▶ pH	A figure expressing the acidity or alkalinity of a solution on a logarithmic scale on which 7 is neutral, lower values are more acid and higher values more alkaline
	▶ Nutrients	Include nitrates and phosphate
	▶ Odour	Distinctive smell, especially unpleasant
	▶ Taste	The sensation of flavour perceived in the mouth
Biological characteristics	▶ Presence of bacteria	For example: E. coli, Legionella, Leptospira, Salmonella, Vibrio cholera, Typhoid, Para-Typhoid, etc.
	▶ Presence of viruses	For example: Hepatitis A, Hepatitis E, Noroviruses, etc.
	▶ Presence of parasites	For example: Protozoa and Helminthes

(Continued)

Table 5.1 (Continued)

Type	Factor	Description
Physical characteristics	▶ Turbidity	Due to suspended solids
	▶ Colour	Its clarity
	▶ Surface characteristics	For example: floating films, foam, etc.
	▶ Temperature	Cool water is generally more palatable than warm water and temperature will have an impact on the acceptability of a number of other inorganic constituents and chemical contaminants that may affect taste. High water temperature enhances the growth of microorganisms and may increase problems related to taste, odour, colour and corrosion
	▶ Radiation	For example: gross alpha and gross beta radiation activity

Heavy metals are of concern because of their effects on the environment and human health:

▶ cadmium (Cd);
▶ chromium (Cr);
▶ copper (Cu);
▶ lead (Pb);
▶ mercury (Hg);
▶ nickel (Ni);
▶ tin (Sn); and
▶ zinc (Zn).

pH scale

The 'p' stands for 'potenz' (German for power/potential to be). The 'H' stands for hydrogen.

The pH of distilled water is 7;

▶ pH 7 is neutral;
▶ any solution with a pH below 7 is an acid; and
▶ any solution with a pH above 7 is an alkali.

Acid rain has a pH of 5.0 or less.

Most acid deposition ranges from pH 4.3 to 5.0 – somewhere between the acidity of orange juice and black coffee.

Water quality objectives and water quality standards for surface water

Water Quality Objectives (WQO) are used to classify water according to actual or potential use.

▶ WQO is a system for classifying waters according to their actual or potential use deriving from a range of EU Directives.

▶ Objectives set down limits, relevant substances or quality criteria based on the use of the specific water.

▶ There are EU Directives specifying what objectives should apply for specified uses, such as bathing water, drinking water, a habitat for shellfish, etc.

Water Quality Standards (WQS) set down limits for indicators of poor water quality, such as level of hazardous substances.

▶ WQS derive from the EC Directive on pollution caused by certain dangerous substances.

▶ Commonly referred to as Black List and Grey List, or List I and List II substances.

▶ Set down limits in concentration of specified dangerous substances allowed in discharges to controlled waters.

The water cycle; sources of water such as groundwater, surface water and desalination

The water cycle describes the continuous movement of water on, above and below the surface of the Earth.

▶ Water is evaporated by solar energy from open waters (seas, rivers and lakes).

▶ Plants emit water vapour into the atmosphere by transpiration and animals by respiration.

▶ Water vapour forms clouds in the high atmosphere where it cools and condenses into rain precipitation, ice or snow and falls back to Earth as rain, hail or snow.

▶ Water returns directly to bodies of water or is absorbed into the ground.

▶ Groundwater sources eventually find their way back into major water body surface runoff.

▶ Water is stored in ice and glaciers.

▶ An essential role of the cycle is to clean and purify water.

The water table, also called the groundwater table, is the upper level of an underground surface in which the soil or rocks are permanently saturated with water. The water table is the top level of groundwater. Surface water is an exposed part of the water table.

Water for agriculture and industry

The abstraction (removal) of surface and groundwater has long been an important source of process water for agriculture and industry across the world. Worldwide agriculture accounts for 70 per cent of all water consumption, compared with 20 per cent for industry and 10 per cent for domestic use.

Impact of water pollution on wildlife

The impact of water pollution on wildlife is wide ranging. The World Wildlife Fund (WWF) states that

> [w]ater becomes polluted from toxic substances dumped or washed into streams and waterways and the discharge of sewage and industrial waste. These pollutants come in many forms – organic, inorganic, even radioactive – and can make life difficult, if not impossible, for humans, animals and other organisms alike.

Water pollution can harm wildlife by:

▶ potential direct toxic effects of some pollutants;
▶ impact from nutrients;
▶ impact from sediments;
▶ impact from suspended solids;

- ▶ impact from oil;
- ▶ consequences of raised temperatures;
- ▶ consequences of depleted oxygen concentrations;
- ▶ impact on crops;
- ▶ impact on reproductive cycles of wildlife;
- ▶ bioaccumulation in the food chain; and
- ▶ damage to biodiversity.

Over-abstraction; water conservation; balancing the different needs

Simple measures can be taken to reduce the amount of water used. The following table identifies the top nine uses of water at home and outlines ways to reduce the amount of water used.

Table 5.2 Uses of water at home

Activity	How much water is used?
Running the tap	8–12 litres per minute
Washing up in the sink	6–8 litres
Washing hands and face	3–9 litres
Taking a normal shower	6–12 litres per minute
Taking a power shower	13–22 litres per minute
Flushing the toilet	6–12 litres
Running a modern dishwasher	15 litres
Running a modern washing machine	60–80 litres
Having a bath	75–90 litres
Using a hosepipe	550–1,000 litres per hour
Making food and drink	6–10 litres

Study point

This is for illustration purposes only, you would not be expected to remember this table.

Source: Anglian Water

The potential effects of pollution on water quality

The four key effects of pollution on water quality are eutrophication; oxygen depletion; silt deposition; and acidification.

Eutrophication

▶ Is an excess of nutrients (mainly nitrogen and phosphorous).
▶ Excess of nutrients is caused mainly by runoff of fertilisers and other discharges containing nutrients, including sewage.
▶ Excess of nutrients leads to algal blooms and excessive plant growth that limit light penetration, deplete oxygen and smother surfaces.
▶ Lack of oxygen may lead to fish dying and decay of plant and animal matter, which then causes anaerobic conditions to develop.
▶ Anaerobic conditions may lead to the release of toxic sulphides into the water.

Oxygen depletion

▶ Is a lowering of dissolved oxygen which can be caused by discharges of effluents containing high levels of biological oxygen demand (BOD) or chemical oxygen demand (COD), or through the decay of organic matter, for example, the polluting potential of milk and orange juice, which both have high levels of BOD.
▶ Lowering of oxygen leads to the death of invertebrates and fish.
▶ May be one of the effects arising from eutrophication.

Silt deposition

▶ Is due to the settlement of fine solids that are discharged or washed into a watercourse from surrounding land.
▶ Occurs during disturbance associated with earthworks, or through the disturbance of natural fine silts on the bed of watercourses.

▶ Consequent smothering of vegetation by silt deposits inhibits photosynthesis, kills invertebrates and affects fish and their spawning grounds.
▶ Accumulation of silts may reduce channel depth.

Acidification

▶ Is due to an increase in acidity of water following discharge of acidic effluents or from acid rainfall.
▶ Acidity may mobilise toxic metals, such as aluminium.
▶ May adversely affect aquatic systems possibly harming or killing fish and invertebrates.

Revision exercise

Describe the essential features and mechanism of the 'water cycle'. (8)

Exam tip

Identify how your organisation can take simple measures to reduce the amount of water that is used.

5.2 Outline the main sources of water pollution

Key revision points

▶ Surface water drainage and risks of contamination from spills, etc.
▶ Drainage from mining, quarrying and ore processing
▶ Process water, sewage and cooling water, leakage from disused process facilities, tanks, etc.

> ▶ Contamination from natural minerals
> ▶ Groundwater: spillage onto unmade ground allowing buildup and seepage through the ground to controlled waters
> ▶ Solids such as grit from sites, plastics, etc.

Surface water drainage and risks of contamination from spills, etc.

There are two main types of drainage systems: surface water drains and foul water drains. Surface water drains carry clean, uncontaminated rainwater from roofs, roadways and hard-standings, etc. Foul water drains carry dirty/contaminated water to sewage treatment works for processing before being returned to a river or sea. Drains are often colour coded: blue for storm and red for foul.

Drainage from mining, quarrying and ore processing

Excavating coal, metal ores or other minerals may expose rocks and minerals that contain sulphur. If sulphur mixes with water and oxygen it can cause groundwater to become acidic, i.e., Acid Mine Drainage (AMD). This can contain highly polluting metals such as lead, zinc, iron, mercury and cadmium. Contaminated water from spoil heaps and stockpiles of crushed and waste rock can also create an acidic discharge called Acid Rock Drainage (ARD).

Process water, sewage and cooling water, leakage from disused process facilities, tanks, etc.

Sources of water pollution from industrial sites include:

- ▶ surface water drainage contaminated by spills;
- ▶ sewage;
- ▶ cooling water (if too hot);
- ▶ effluent discharges;
- ▶ wrong connections;

- inappropriate disposal of waste;
- leaks from operational or disused equipment;
- water spread onto hard standing surfaces from washing operations can find its way into streams via surface water drains;
- cross connections of contaminated drains to surface water drains;
- badly managed waste facilities and wash off of solids;
- in emergencies, accidental spills or firewater caused by pollution unless contained; and
- leachate from landfill sites.

The effects caused by discharge of water pollutants into surface waters include:

- oxygen depletion;
- toxicity and ecotoxicity to plants, aquatic organisms, animals, birds or people;
- eutrophication associated with nutrients;
- infection risks associated with pathogenic substances;
- sedimentation leading to changes in flow;
- reduced light penetration and harm to plants;
- coating of feathers on birds from floating contaminants;
- potential for bioaccumulation and possible interactions;
- reactions with other pollutants; and
- nuisance effects and the impact on downstream users.

Contamination from natural minerals

There is a range of natural sources of pollution. These include:

- volcanic eruptions;
- algae blooms (rapid increase or accumulation in the population of algae);
- radon gas – in various parts of the world, including UK, Spain, Romania, etc.
- arsenic – toxic levels have been found in drinking water in Bangladesh: up to 77 million people in Bangladesh have been exposed to toxic levels of arsenic from drinking water in recent decades; and
- fluoride – toxic levels have been found in drinking water in India and China.

Groundwater: spillage onto unmade ground allowing buildup and seepage through the ground to controlled waters

Water is present almost everywhere underground, but some geological formations are impermeable:

▶ meaning that water can hardly flow through them; and
▶ some are permeable – they contain fine holes that allow water to flow.

Permeable formations that contain groundwater are known as aquifers. Holes that water flows through can be spaces between individual grains in a rock like sandstone, or they can be networks of fine cracks.

Solids such as grit from sites, plastics, etc.

Wastewater contains grit, plastics, etc. that can interfere with water treatment processes. A screening process will be used to remove coarse then fine objects to prevent damage to equipment. Various techniques can be used to remove objects including:

▶ aerated grit chambers;
▶ vortex-type grit chambers;
▶ detritus tanks; and
▶ hydrocyclones.

Revision exercise

(a) A spillage of a hazardous substance is an example of a source of water pollution. **Outline FOUR** additional sources of water pollution. (4)
   Part b) is covered in 5.3.

(b) **Outline** measures to reduce the risk of spillage of hazardous substances in a workplace. (4)

Exam tip

Identify the potential sources of water pollution from your organisation.

5.3 Outline the main control measures that are available to reduce contamination of water sources

Key revision points

▶ Control hierarchy: eliminate, minimise, render harmless, with examples
▶ Control methods
▶ Controls for storage and spillage
▶ Controls for waste water
▶ Difficulties in funding and maintaining equipment in some countries or environments

Control hierarchy

The control hierarchy to reduce contamination of water sources follows: eliminate, minimise and finally render harmless.

▶ **Eliminate/prevent pollution** by ceasing or changing the process or the materials used. For example:
  ▷ Eliminating the use of chemicals that will cause pollution.
  ▷ Remove lead pipes, etc.
▶ **Minimise** to the lowest practicable level by means of control devices, process management or choice of materials. For example:
  ▷ Improve safe systems of work, information, instruction, training and supervision.
  ▷ Ensure bunding is sufficient around storage tanks, etc.

▶ **Render harmless** by reducing the polluting content. For example:
  ▷ Treat wastewater in accordance with national and international standards/legislation.

The principle of the hierarchy of control is to address the **source** of the water pollution first. If this cannot be achieved or does not resolve the problem, then an attempt should be made to control along the transmission **path**. If this is effective, then the **receiver** is not affected.

Table 5.3 Source–pathway–receptor example

Source	Pathway	Receptor/Receiver
A spill on ground from ▶ Oil drums ▶ Tanks	▶ Drains ▶ Through soil ▶ Over hard surfaces	▶ River ▶ Stream ▶ Lake ▶ Groundwater ▶ Flora and fauna ▶ People

Control methods

▶ **Permits to discharge**: Some national and international standards/legislation require a permit to be in place to allow companies to discharge into the watercourses.

▶ **Monitoring water quality**

Three common measurements used are:

▶ **Chemical Oxygen Demand (COD)** used to measure the amount of organic pollutants found in water.

▶ **Biological Oxygen Demand (BOD)** used to measure how fast biological organisms use up oxygen in water.

▶ **Total Oxygen Demand (TOD)** used to measure the total amount of oxygen necessary for the complete oxidation of organic and inorganic compounds present in water.

Controls for storage and spillage

Basic control measures will include:

▶ preventing any spillages occurring (through operating procedures, maintenance, etc.);

Basic control measures will include:

▶ screening and solids separation;
▶ sedimentation;
▶ floatation (solids float to surface buoyed up by injected air);
▶ filtration;
▶ centrifugal separation; and
▶ correction of pH.

Difficulties in funding and maintaining equipment in some countries or environments

Bunded tanks, separators and water treatment systems can be expensive to purchase initially. Further costs would include the installation and maintenance of the equipment. General day-to-day running costs of this equipment will be higher in harsh environments – for example, very hot regions. This may become too expensive in some developing countries. If the correct equipment is not used and maintained correctly, this could cause harm to the workers and the local environment through water pollution.

Revision exercise

(a) **Give** the meaning of 'safe drinking water'. (2)
(b) **Give** the meaning of the following water quality indicators:
  (i) the Chemical Oxygen Demand (COD) (2)
  (ii) the Biochemical Oxygen Demand (BOD) (2)
  (iii) the Total Oxygen Demand (TOD). (2)

Exam tip

Don't get bogged down with the detail!

EC 1.6

Control of waste and land use

Learning outcomes

6.1 Outline the significance of different waste categories and the relationship between category and route of disposal ☐

6.2 Explain the importance of minimising waste ☐

6.3 Outline how to manage waste ☐

6.4 Describe outlets available for waste ☐

6.5 Outline the risks associated with contaminated land ☐

6.1 Outline the significance of different waste categories and the relationship between category and route of disposal

Hazardous and non-hazardous types of waste

Waste Framework Directive (Directive 2008/98/EC) defines waste as *'any substance or object which the holder discards or intends or is required to discard'*.

▶ Waste means any substances or objects which the holder discards or intends or is required to discard.
▶ Defined within Environmental Protection Act 1990 (applies to the UK).

Wastes will fall into one of three categories. Those that:

▶ are always hazardous: e.g., lead, acid batteries or fluorescent tubes;
▶ may or may not be hazardous and need to be assessed: e.g., ink or paint; and
▶ are never hazardous: e.g., edible oil.

Some types of waste are harmful to human health, or to the environment, either immediately or over an extended period of time. These are called hazardous wastes.

Other main categories

National and international legislation uses specific definitions for waste including:

Table 6.1 Types of wastes

Types of wastes	Description
Hazardous	Are those wastes that are harmful to human health, or to the environment, either immediately or over an extended period of time. For example, materials that are toxic, corrosive, etc. Note this also includes asbestos
Non-hazardous	Includes paper and food waste. That is anything that is not harmful to human health, or to the environment, either immediately or over an extended period of time

Table 6.2 Categories of wastes

Category of waste	Description
Inert waste	Are those that do not react or decompose and are stable when landfilled, for example, bricks, concrete
Clinical waste	Is any waste which consists wholly, or partly, of: (a) human or animal tissue (b) blood or other body fluids/excretions (c) drugs or other pharmaceutical products (d) swabs or dressings (e) syringes, needles or other sharp instruments Waste which, unless rendered safe, may prove hazardous to any person coming into contact with it Any other waste arising from medical, nursing, dental, veterinary, pharmaceutical or similar practice, investigation, treatment, care, teaching or research, or the collection of blood for transfusion
Radioactive waste	Is any material that is either radioactive itself or is contaminated by radioactivity, for which no further

(Continued)

Table 6.2 (Continued)

Category of waste	Description
	use is envisaged. Radioactive waste has three different categories:
	▶ High-Level Waste (HLW)
	▶ Intermediate-Level Waste (ILW)
	▶ Very Low-Level Waste (VLLW)
Controlled waste	Is a term used in UK law for any type of waste from households, industrial premises, commercial, etc. mining and agricultural waste. This includes
	1 **Household waste** – from domestic property, from charities, royal palaces, moored houseboats, litter collections, etc.
	2 **Industrial wastes** – from laboratories, workshops, vehicles not used for domestic purposes, etc.
	3 **Commercial wastes** – from offices, hotels, courts or government departments
	4 **Mining waste** – from mines
	5 **Agricultural waste** – from farms
Special waste	Is a term used in Scotland. It is essentially any waste with hazardous properties which may render it harmful to human health or the environment
Electrical and electronic equipment waste	Equipment that includes electrical and electronic components which also includes batteries
	Waste Electric and Electronic Equipment (WEEE)
	Hazardous WEEE
Biodegradable waste	Waste that can undergo bacterial decomposition

Revision exercise

Outline the meaning of the following terms:

(a) controlled waste (2)
(b) inert waste (2)
(c) non-hazardous waste (2)
(d) hazardous waste (2)

Exam tip

Look at the categories of waste and apply them to a named waste policy.

6.2 Explain the importance of minimising waste

Key revision points

▶ The problems of waste disposal due to increased volumes from growing populations and higher standards of living
▶ The waste hierarchy
▶ Benefits and limitations of recycling
▶ Barriers to re-use and recycling and how they can be overcome

The problems of waste disposal

Due to increased higher standards of living and from growing populations (currently, it is estimated the world population is approximately 7 billion people; by 2030 this is expected to be above 8 billion and by 2050 this is expected to be above 9 billion), the amount of waste that is produced is increasing exponentially. As more people are on the earth, they are producing more waste.

The cost of disposal is increasing, due to more waste being produced and a limited number of landfill sites available.

Burying waste at landfill sites increases the risk of waste polluting the water table; release of methane (a greenhouse gas); spread of disease by rats and other vermin; and nuisance problems: i.e., odour, noise, dust, etc.

The waste hierarchy

The following table identifies each stage and outlines practical examples of the waste hierarchy.

Table 6.3 Waste hierarchy

Waste hierarchy	Meaning
Prevent	Improving the durability, to make the products last longer
Minimisation	Reducing the amount of waste produced i.e., buying only the amount of product needed so there is minimal leftover which will become waste
Reduce	Using fewer materials in the products
Re-use	Repair and refurbishment of the products to be recycled in their current form
Recover	Recycling the material in products, to be used in new products/recovering energy – i.e., using products as fuel
Dispose	Sending products to landfill, or incineration, without energy recovery

The following list identifies the tangible and intangible benefits of minimising waste.

- ▶ Tangible:
    - ▷ more efficient use of materials;
    - ▷ more efficient equipment and lighting;
    - ▷ less waste to segregate;
    - ▷ less waste to store prior to disposal;
    - ▷ saving money through purchasing materials – but don't buy in bulk to save when most end up as waste; and
    - ▷ saving money through less waste disposal charges and taxes.
- ▶ Intangible:
    - ▷ enhancing staff morale;
    - ▷ raising company profile with external stakeholders and improving public relations; and
    - ▷ helping to ensure compliance with legislation.

Benefits and limitations of recycling

Benefits and limitations of recycling have been identified in the table below.

Table 6.4 Benefits and limitations of recycling

Benefits of recycling	Limitations of recycling
▶ Assist with legal compliance	▶ Limited space
▶ Reduced demand for raw materials	▶ Unsuitable containers
▶ Reduced emissions and water use and saving energy associated with new products	▶ Lack of knowledge
	▶ Attitudes that affect behaviour
▶ Reduced damage due to extraction of resources	▶ Recycling can take up time and requires some established processes and leadership to make it work
▶ Less waste to landfill	
▶ Fewer greenhouse gas emissions	▶ Initial costs involved
▶ Conserve habitats and wildlife	▶ Needs to be a market for recycled materials and goods
▶ Reduced costs for materials	
▶ Reduced costs for waste disposal	▶ This market can be influenced by the perceived quality of recycled products or goods made from recycled materials
▶ Creates jobs	
▶ Improves public image	

Barriers to re-use and recycling and how they can be overcome

To enable re-use and recycling of materials/products, an organisation will need to ensure it has the right equipment. Management and the workforce will need to have the correct information and advice – for example, this can be achieved in the UK working with organisations such as the National Industrial Symbiosis Programme (NISP).

Practical measures include:

▶ keeping it simple will help improve recycling rates;
▶ markets for recycled materials or goods can be developed;
▶ provision of information and advice;
▶ provision of the right equipment;

▶ recycling can be included in contracts with suppliers;
▶ strong policy and leadership with demonstrable benefits will reinforce the message;
▶ training;
▶ working with customers directly; and
▶ working with suppliers.

Where possible, keeping the re-use and recycling systems simple will help improve recycling rates.

In addition to the cost savings of waste disposal, recycled products may produce a new market stream. It is possible to develop new contracts that include re-use and recycling systems for suppliers, etc.

Having a strong policy and leadership with demonstrable benefits will reinforce the organisation's commitment to the environment.

Revision exercise

(a) **Outline** the business benefits of minimising waste. (5)
(b) **Identify** the environmental options, in hierarchical order, for the waste hierarchy. (3)

Exam tip

Identify ways in which your organisation could increase its recycling.

6.3 Outline how to manage waste

Key revision points

▶ Recognition of the key steps: on-site separation, storage, transportation and disposal

▶ Responsible waste management
▶ Segregation, identification and labelling
▶ Packaging waste
▶ Electrical and electronic waste
▶ Regulatory documentation
▶ Differing requirements for domestic/commercial/industrial waste in many countries
▶ Waste from construction projects

Recognition of the key steps: on-site separation, storage, transportation and disposal

Table 6.5 Key steps that a producer of waste should take

Key steps	Description
Separation/ segregation	The process of identification and labelling of waste material should be undertaken to ensure that non-compatible waste is not stored together – for example, waste that could chemically react and cause a fire or release of poisonous gases
Storage	Different types of waste should be stored in suitable containers – for example, liquid waste should not be able to leak, combustible materials should be stored away from buildings, etc.
	Different wastes should be stored separately (metal, wood, paper, etc.) or different chemicals, i.e., flammable and oxidisers
Transportation	Those collecting waste must comply with local national legislation. It will be the legal requirement of the waste producer to check to make sure that their waste carrier is able to transport the waste; must have a waste carrier's licence
	Note: Hazardous waste needs to be disposed of in UN approved containers
Disposal	Will include sending products to landfill, or incineration

Responsible waste management

Waste holders are those who are producers, transporters or disposers of waste and therefore have a duty of care. Their responsibilities should include:

▶ waste should be classified; that is identified as either hazardous or non-hazardous;

▶ waste should be stored safely and securely;

▶ laws/guidelines for moving waste off business premises should be followed; and

▶ checks should be made to ensure that an authorised waste carrier carries out the transportation of waste.

The premises that store hazardous waste may need to be registered and a permit may be required to store, treat, transport or dispose of any waste.

All waste producers should undertake a duty of care and ensure:

▶ that waste does not cause harm to people or the environment;

▶ security and protection of waste from weather and vermin;

▶ segregation of different types of waste and prevention of escapes;

▶ labelling, written descriptions and the use of transfer and consignment notes provide a paper trail;

▶ disposal has been by means of registered persons and to authorised sites; and

▶ training, supervision and inspection ensure correct treatment of waste.

Segregation, identification and labelling

The first of the key steps is separation/segregation. The following outlines some specific actions that might be required.

▶ Proper labelling and site management, including record keeping, will be required.

▶ Supervision and site inspections need to be carried out regularly.

▶ In the event of some unplanned incident, it is also necessary to have emergency plans in place.

▶ If stored outside, containers deteriorate, i.e., plastic – in sunlight, due to UV light.

Packaging waste

Packaging is defined as anything that is used to contain and protect raw materials or a product and is thrown away after the product is consumed. Manufacturers are encouraged, and in some countries are required, by local laws to reduce the amount of packaging that is used.

▶ Packaging is anything that is used to contain and protect raw materials or a product and is thrown away after the product is consumed.

▶ More than 10 million tonnes of packaging waste arises in the UK from industry, commerce and households.

▶ More than 6 million tonnes of packing waste are accepted for recovery either in the UK or overseas.

▶ Recovery of waste is good for the environment and helps the UK meet its recovery and recycling targets.

▶ Good for business as those that cut down on waste packaging can save considerable amounts of money.

Factors to consider with the storage of waste include:

▶ Segregation of waste and containment to prevent escape especially choosing suitable containers are key measures.

▶ Liquid wastes need some bunding arrangements.

▶ All wastes may need secure sites and protection from the weather, pests and scavengers.

▶ Incompatibility risks of different materials and location of sites away from buildings, watercourses or other potential sites where damage may be caused need to be considered.

Electrical and electronic waste

Waste Electrical and Electronic Equipment Directive (WEEE Directive) is the European Community directive 2002/96/EC. The recast WEEE Directive 2012/19/EU was published on 24 July 2012.

Categories of electrical and electronic equipment waste under the directive are:

- large household equipment;
- small household equipment;
- IT and telecommunications;
- consumer equipment;
- lighting equipment;
- electrical and electronic tools;
- toys, leisure and sports equipment;
- medical devices;
- monitoring and control devices; and
- automatic dispensers.

The above items should not be disposed of as normal household/commercial waste but should be disposed of at a specialist waste facility.

Regulatory documentation

In the UK the following legislation requires companies to keep specific documentation:

- Hazardous Waste Regulations 2005
- Environmental Protection Act 1990
- Environmental Permitting (England and Wales) Regulations 2016

Differing requirements for domestic/commercial/industrial waste in many countries

There are differing requirements for domestic, commercial, industrial and construction waste in many countries. Environmental managers/

practitioners will need to review their own countries' legislation and guidance.

Waste from construction projects

The range of waste from construction products is vast. The following list identifies some of the common waste materials that may result from a construction project:

▶ architectural fixtures;
▶ asphalt;
▶ brick, tile and masonry materials;
▶ carpet;
▶ concrete and concrete blocks;
▶ electric waste (wires, etc.);
▶ ferrous metal (steel, iron, etc.);
▶ glass;
▶ gypsum wallboard scrap;
▶ insulation;
▶ non-ferrous metals (copper, aluminium, etc.);
▶ paper and cardboard;
▶ plumbing fixtures;
▶ plywood, oriented stand board and plaster board;
▶ rigid foam;
▶ tree, shrubs, soil;
▶ tile and masonry materials; and
▶ untreated wood.

Many of the above can be re-used or recycled.

Revision exercise

Identify EIGHT categories of electrical and electronic equipment waste under the WEEE Directive 2012/19/EU. (8)

> **Exam tip**
>
> To help you understand the key steps of waste management
> (above), consider how waste is managed at your place of work.

6.4 Describe outlets available for waste

> **Key revision points**
>
> ▶ Landfill and incineration as ultimate disposal routes;
>   advantages and disadvantages
> ▶ Domestic waste sites, waste transfer stations, waste
>   treatment facilities involving recovery operations
> ▶ Waste disposal in developing countries and problems
>   associated with domestic waste
> ▶ Costs and the impact of landfill and aggregate taxes

Landfill and incineration as ultimate disposal routes; advantages and disadvantages

Landfill

The major potential problems (disadvantages) concerned with landfill
sites include the production of landfill gas (mix of different gases,
including methane) and leachate (water that has percolated through
a solid and leached out some of the constituents) and causing a
nuisance (from odour, dust, noise, vermin, etc.). In addition, waste
takes time to break down, and therefore it remains an environmental
concern for a significant length of time. Another issue is the
increasing costs imposed on landfills.

The advantages/benefits of landfill include the fact that it is a cheap form of disposal (in comparison with incineration). Well-designed and managed sites can reduce the risk to the environment and humans. It also enables waste to be disposed of locally, thereby reducing the transportation of waste. Landfill sites can be used to fill in disused quarries, etc.

Incineration

The major potential problems (disadvantages) concerned with incineration are potential risk to the environment and humans, i.e., through air pollution. Some of the materials incinerated could have been recycled – for example, plastic. The residual ash left over still needs to be disposed of in a landfill.

The advantages/benefits of incineration include the fact that it is the only option for the disposal of medical waste and some other hazardous wastes. The overall amount of residual waste left after incineration is, however, greatly reduced compared with its initial state.

Domestic waste sites, waste transfer stations, waste treatment facilities involving recovery operations

Civic amenity sites are those used for the **disposal of domestic waste**. Where practicable, waste will be separated into:

▶ paper/cardboard;
▶ glass;
▶ plastics;
▶ metal; and
▶ hazardous wastes (e.g., motor oil, batteries).

The temporary storage of waste, prior to its collection, is stored at a **waste transfer station**. These sites will store waste before it is transported to another location for further treatment or disposal.

Waste treatment facilities will use a variety of techniques to sort and remove recyclable materials. The remaining waste is taken to landfill sites or used in the generation of energy (i.e., through incineration).

Waste disposal in developing countries and problems associated with domestic waste

Developing countries produce lower levels of waste per capita with a higher proportion of organic material in the municipal solid waste stream.

If measured by weight, organic (biodegradable) residue constitutes at least 50 per cent of waste in developing countries.

Labour costs are relatively low but waste management is generally a higher proportion of municipal expenditure.

As urbanisation continues, municipal solid waste grows faster than urban populations because of increasing consumption and shortening product life spans.

Costs and the impact of landfill and aggregate taxes

Various countries will charge companies for the disposal of their waste in the form of **landfill taxes.**

These financial payments have advantages and disadvantages.

Advantages include:

- ▶ alternate methods are considered (see waste hierarchy below) by companies;
- ▶ reduces the need for more raw materials;
- ▶ suppliers are reducing the packaging of the products; and
- ▶ more lean manufacturing processes are being developed to reduce wastage.

Disadvantages include:

- ▶ increasing the risk of fly-tipping (dumping waste).

Various countries will charge companies for extraction of minerals in the form of **aggregate taxes**. The primary purpose is to raise revenue for the government.

Revision exercise

(a) **Identify** the categories of waste. (4)
(b) **Outline**:
　(i) the main advantages (2)
　(ii) the main disadvantages (2)

associated with the landfilling of wastes.

Exam tip

Identify the current charges for disposal of waste at your organisation.

6.5 Outline the risks associated with contaminated land

Key revision points

▶ The potential effects of contaminated land to the environment
▶ Liabilities of an organisation from contaminated land

The potential effects of contaminated land to the environment

Contaminated land is caused by ground or surface water pollution by leached contaminants from waste sites. Also leakage from old oil drums, pipes, etc. will contaminate the land. Contamination of water sources (for example: through deliberate pouring of harmful liquids) will also have a negative impact on the land. Air pollution will also have a negative impact on the land.

The potential effects of contaminated land on the environment include:

▶ Ground or surface water pollution by leached contaminants.
▶ Human health may be affected by:
  ▷ direct contact with the soil;
  ▷ ingestion; and
  ▷ inhalation of dust or vapours.
▶ Odours or other forms of air pollution.
▶ Affect wildlife directly, or by loss of habitat.
▶ Fire or explosion risks, due to building up of methane, etc.
▶ Damage to building materials.
▶ Radiation risk.
▶ Loss of amenity value.
▶ Restrictions on future use without some form of treatment.

Liabilities of an organisation from contaminated land

Contaminated land can be defined as:

Any land in such condition, by reason of substances in, on or under the land, that presents a significant possibility of significant harm being caused or pollution of controlled waters being or likely to be a cause of harm to people or other receptors.

An organisation may be held responsible for any damage it has caused that contaminates land. It may be fined or be required to undertake remediation action, i.e., to put the land back to how it was prior to the pollution occurring.

Revision Exercise

(a) **Identify** the main ways in which contaminated land can pollute the environment. (4)
(b) **Outline** the potential environmental effects that may arise from contaminated land. (4)

Exam tip

Identify ways in which contaminated land could be caused by your organisation – what can be done to prevent this?

EC 1.7

Sources and use of energy and energy efficiency

Learning outcomes	
7.1 Outline the benefits and limitations of fossil fuels	☐
7.2 Outline alternative sources of energy and their benefits and limitations	☐
7.3 Explain why energy efficiency is important to the business	☐
7.4 Outline the control measures available to enable energy efficiency	☐

7.1 Outline the benefits and limitations of fossil fuels

> **Key revision points**
>
> ▶ Examples of fossil fuels
> ▶ Benefits and limitations of their use as an energy source

Examples of fossil fuels

Different types of fossil fuels (gas, oil and coal) were formed in the ground depending on what combination of animal and plant debris was present, how long the material was buried and what conditions of temperature and pressure existed when they were decomposing. Coal is the product of dead plants that have been buried and compacted beneath sediments, whereas crude oil and natural gas are the products of the deep burial and decomposition of dead plants and animals.

Fossil fuels are described as non-renewable:

▶ natural gas (is formed primarily of methane);
▶ crude oil (which is a dark oil consisting mainly of hydrocarbons, i.e., an organic compound consisting entirely of hydrogen and carbon); and
▶ coal (which is composed of carbon, hydrogen and oxygen).

Fossil fuels will also contain trace amounts of other chemicals, for example, sulphur – therefore when burned will release the gas sulphur dioxide.

With sufficient levels of oxygen, the burning of fossil fuels (carbon and hydrogen) will produce carbon dioxide and water vapour. In the event of insufficient levels of oxygen and the burning of fossil fuels carbon monoxide will be produced.

Gas and oil need to be extracted from the earth before they can be processed to produce energy.

The components in crude oil can be separated by fractural distillation. The crude oil is evaporated and its vapours are allowed to condense

136

at different temperatures in a fractionating column. The crude oil is split into fractions containing similar sized molecules. The lower the number of carbon atoms present in the molecules results in different physical states (i.e., gases, liquids and solids):

▶ Gases (methane, ethane, propane and butane).
▶ Liquids (solvents, petrol, diesel, heating oil, motor oil, etc.).
▶ Solids (lubricants, candles, road paving [bitumen] and industrial fuel for steel production [coke]).

Coal needs to be extracted from mines (colliery) or open cast/cut. It is then transported to coal-powered stations and converted into electrical energy.

▶ Gas-fired power stations are more efficient than coal-fired power stations.
▶ Gas-fired power stations are about 50 per cent efficient.
▶ Coal-fired power stations are about 30–35 per cent efficient.

Atmospheric pollutants that may be released from a coal-fired power station include:

▶ sulphur and nitrogen oxides;
▶ carbon dioxide and carbon monoxide;
▶ dust or fly ash;
▶ smoke or soot; and
▶ metals.

Benefits and limitations of their use as an energy source

Table 7.1 Fossil fuels

	Benefits	Limitations
	1 Generation of energy is simple/straightforward process	1 Causes damage to the environment (i.e., acid rain)
	2 Inexpensive to extract and turn into energy	2 Causes ill health (i.e., poor air quality)

(Continued)

Table 7.1 (Continued)

Benefits	Limitations
3 Power stations can be built almost anywhere	3 Factor of climate change
4 Fossil fuels can be easily transported	4 Non-sustainable in long term (i.e., they will run out)
5 A convenient form of energy source, i.e., for vehicles, etc.	5 Can be dangerous to extract and turn into energy – this may result in injury or ill health to those involved with the process
6 They have a high calorific value, i.e., the energy content is high	6 Concerns over security of supply – causing prices to significantly fluctuate

Revision exercise

Outline:

(i) the benefits (4)
(ii) the limitations (4)

associated with the use of fossil fuels as an energy source.

Exam tip

Identify what types of fossil fuels are used by your organisation. Is it practicable to use alternative sources of energy?

7.2 Outline alternative sources of energy and their benefits and limitations

Key revision points

▶ Solar, the wind, hydroelectric, wave and tidal power, geothermal, nuclear, combined heat and power (CHP), bio-digesters, methane recovery

▶ Benefits and limitations of their use in each case
▶ Problems of energy supply in remote regions and developing countries

Solar, the wind, hydroelectric, wave and tidal power, geothermal, nuclear, combined heat and power (CHP), bio-digesters, methane recovery

There is a wide range of alternative sources of energy, including: solar, the wind, hydroelectric, wave and tidal power, geothermal, nuclear, combined heat and power (CHP), bio-digesters and methane recovery. Each source has its own benefits and limitations.

Key advantages for using alternative energy sources include:

▶ reduce dependency on fossil fuels;
▶ reduce greenhouse gas emissions;
▶ reduce sulphur oxide and nitrogen oxide emissions;
▶ reduce costs of producing energy over time (especially if supported by grants or other incentives);
▶ energy security;
▶ lower pollution caused by the extraction or distribution of fossil fuels; and
▶ opportunities for new businesses and job creation.

Solar

Solar energy is a renewable form of energy. A small percentage of the sun's energy is intercepted by the Earth's atmosphere.

Solar power can be captured via solar collectors to produce hot water for washing or space heating in buildings. Such collectors are in widespread use in sunny countries.

There are two main processes for harnessing solar energy directly and converting it into electrical energy: Solar Thermal Electric (STE) (also known as Concentrated Solar Power [CSP]) and Solar Photovoltaics (SPVs).

139

Some studies suggest CSP sites in Southern Europe and North Africa could generate enough electricity to replace all of Europe's nuclear power and vastly reduce electricity consumption of fossil fuels. Covering just 1 per cent of the world's deserts with CSP could produce enough electricity to meet global demand.

SPVs use semiconducting materials, such as silicon, that produce electricity when exposed to daylight. SPV units are placed on roofs of buildings and can produce enough electricity for all, or a portion, of the building's needs. SPVs can be used on a range of other applications: from charging mobile phones to powering satellites.

Wind

Windmills have harnessed energy for centuries. The basic principles have been modernised to produce electrical energy.

Wind turbines are a renewable form of energy. They can work on land (onshore) or in the sea (offshore). As air flows from warmer to cooler areas, this results in winds that can be used to power wind turbines to produce electricity.

Wind turbines are one of the most technically and economically developed forms of renewable electricity generation.

Hydroelectric

Hydroelectric power is a renewable form of energy and is generated from rivers, lakes, waterfalls and reservoirs. Hydro-energy has been harnessed since before the turn of the 19th century, i.e., by using water wheels for mills, etc. Electricity, produced through hydroelectric power, is created by flowing water through a turbine. The generation can be on a large or small scale. In most cases, the water will need to have a drop height of at least two metres and be of a large volume. The use of large amounts of hydropower involves the construction of large dams and the flooding of massive areas.

Wave and tidal power

Wave and tidal power are renewable forms of energy. Currently, the generation of electricity from wave and tidal power is limited. The electricity can be generated from the rise and fall of waves or from underwater turbines (similar to wind turbines) that move as a result of the tidal currents.

Geothermal

Geothermal is a renewable form of energy. It is possible to heat water by using the Earth's internal heat. Water needs to be pumped through tubes buried underground that can harness the energy. The resulting hot water, or in some cases steam, can be used to generate electricity or be used to warm water for central heating and hot water.

Nuclear

Nuclear energy is non-renewable, however unlike gas, oil and coal, it is not produced from fossil fuels. Conventional nuclear technology uses fission – splitting atoms of heavy metals such as uranium-235 that will produce the same amount of energy as the combustion of around 3,000 tonnes of coal. The heat generated is used to produce steam to drive turbines that in turn generate electricity.

Deposits of uranium could, possibly, last for several centuries. The release of energy stored in the nuclei of atoms such as uranium-235 produces large quantities of energy. Uranium-235 is an isotope of uranium. An isotope is a form of the same chemical element that has the same number of protons in their nuclei, but a different number of neutrons.

Uranium ore is mined by underground or open-cut methods. The ore is treated with acid to dissolve the uranium, which is then recovered from the solution. Uranium is used in keels of yachts, counterweights for aircraft control surfaces, as well as for radiation shielding!

A major disadvantage of using nuclear power is the potentially devastating effects on people and the environment in the incidence of a catastrophe – e.g., Japan's Fukushima Daiichi nuclear power station on March 11, 2011, caused the most extensive release of radioactivity since the Chernobyl accident in 1986.

Combined Heat and Power (CHP)

Combined Heat and Power (CHP) is the simultaneous generation of usable heat and power (usually electricity) in a single process. CHP power can be used to heat nearby towns and industrial sites.

Bio-digesters and methane recovery

Bio-digesters convert organic waste into biogas and liquid fertiliser. The biogas can then be used for electrical production that is then used mainly for cooking and heat energy. In addition, the waste that is left from the process can be used as a fertiliser. The process is called Anaerobic Digestion (AD). It is essentially a controlled and enclosed version of the anaerobic breakdown of organic waste in a landfill that releases methane. By preventing methane from venting freely into the atmosphere, these systems can help reduce emissions that contribute to climate change. Methane is a potent greenhouse gas that traps heat at twenty-three times the rate of carbon dioxide.

Biogas composition includes:

▶ methane;
▶ carbon dioxide;
▶ nitrogen;
▶ hydrogen;
▶ hydrogen sulphide; and
▶ oxygen.

Benefits and limitations of their use in each case

Table 7.2 Alternative sources of energy

Type	Benefits	Limitations
Solar	▶ Excess electricity can be fed into the grid ▶ Once installed it generally requires little or no maintenance ▶ Can be used as part of building i.e., can replace conventional tiles or cladding	▶ Requires sunlight ▶ Initial high cost ▶ Limited efficiency (older versions) ▶ Requires maintenance ▶ Users have to sell their energy back to the grid ▶ Long term investment before any profit return
Wind	▶ Good method for supplying energy to remote areas ▶ No waste or greenhouse gases produced during generation of electricity ▶ The land around turbines may still be used for agriculture ▶ Wind farms can be tourist attractions ▶ Wind is a free energy source ▶ Zero emissions after less than a year	▶ Not every day will have enough wind to generate electricity ▶ View held, by some, that wind turbines are unsightly ▶ May affect television reception for people living near a turbine ▶ Older versions are known to be noisy ▶ Offshore installation better wind but more difficult construction ▶ May have to be stopped during high winds above 65 mph
Hydroelectric	▶ On a small scale, hydropower does not result in any significantly adverse environmental impacts ▶ The electricity can be used locally or fed into the grid	▶ The use of large amounts of hydropower involves the construction of large dams and the flooding of massive areas ▶ Negative impacts on fish and other wildlife

(*Continued*)

Table 7.2 (Continued)

Type	Benefits	Limitations
		e.g., loss of breeding grounds, etc.
		▶ Increased prevalence of disease in the water
		▶ A decrease in nutrients used in agriculture downstream
		▶ Visual impact on the environment
Wave and tidal power	▶ Low visual and few environmental effects	▶ High costs in development
		▶ Possible impacts on shipping, fishing and marine life
		▶ Visual impacts
Geothermal	▶ Ground source heat can be carbon-neutral	▶ Space is needed outside to lay the pipes underground, which makes it inappropriate for a townhouse with a small garden
	▶ Free and continuous in some places such as Iceland	
		▶ Systems need a dedicated boiler as back-up
Nuclear	▶ No CO_2 emissions occur from its generation	▶ Radioactive waste has to be stored for an indefinite length of time
	▶ Almost endless supply of raw material	
		▶ Concerns about radioactive emissions from normal operations
		▶ Risk of accidental leak (e.g., Fukushima)
		▶ Concerns regarding terrorism threats
Combined Heat and Power (CHP)	▶ Very efficient use of fuel	▶ Needs costly hot water pipeline network
	▶ Can run using waste from other industries and communities	▶ Most energy generation is from gas

Type	Benefits	Limitations
		► CHP more expensive than simple coal and gas generation
Bio-digesters and methane recovery	► Biogas is a sustainable substitute for the propane, kerosene and firewood that many rural families in developing countries use for their domestic energy needs	► Sulphurous compounds can lead to odour
		► Transport
		► Effluvia
	► Use waste, which would have been left to decompose or be sent to landfill	

Problems of energy supply in remote regions and developing countries

In excess of one billion people in the developing world lack access to energy sources such as gas, oil and electricity. Approximately one-third of all energy consumed in the developing world is produced from biomass.

Electricity should not be seen as a luxury good reserved only for rich countries, but a necessity for fulfilling basic human needs. It is estimated that 25 per cent of the world's population uses 75 per cent of the world's energy. Access to electricity:

► Aids water and irrigation through cleansing and pumping and can thus boost agricultural production.
► Provides education by providing light and communication tools.
► Improves productivity, since electricity also allows for irrigation, crop processing, food preservation, water pumping, fencing, agro-processing, ice making, etc.
► Helps to prevent natural disasters by giving the possibility of installing radio communication facilities, enabling remote weather measuring, data acquisition and transmission, etc.

▶ Improves safety measures, for example, street lighting, security lighting, remote alarm systems, electric fences, road signs, railway crossing and signals, warning lights, etc.

▶ Improves gender equality by relieving women of fuel and water collecting tasks.

▶ Reduces child and maternal mortality as well as disease incidence by enabling refrigeration of medication as well as access to other equipment.

▶ Nearly 1.5 billion people in developing countries have no access to clean water.

▶ Nearly 11 million children die each year from preventable diseases; nearly 1.5 million of these deaths are attributable to dirty water and poor hygiene practice.

▶ Reduces isolation and marginalisation by enabling communication, e.g., radio and the Internet.

Issues faced include finding fuel including wood and other traditional fuels such as dung and crop residues. Using other 'traditional fuels' prevents them from being used for fertiliser.

Air pollution from inefficient burners is another major concern – for example, studies have shown that non-smoking women in India and Nepal who have cooked on biomass stoves for many years have a higher-than-normal incidence of chronic respiratory disease.

Revision exercise

Identify FOUR sources of energy other than fossil fuels **AND**, in **EACH** case, **outline** how the energy is generated. (8)

Exam tip

How can alternative sources of energy be used in your organisation?

7.3 Explain why energy efficiency is important to the business

---

Key revision points

▶ Reductions in carbon dioxide emissions
▶ Savings in energy bills and peak load management

Reductions in carbon dioxide emissions

Man-made activities that create/emit carbon dioxide (CO_2) are the combustion of fossil fuels (coal, gas and oil) for generation of electricity and transportation; in addition, certain industrial processes emit CO_2.

A carbon footprint is a measure of how much each person is contributing to the gases that contribute to global climate change. Although we talk about a 'carbon footprint', it would be more accurate to talk about a 'carbon dioxide footprint'. A carbon footprint is normally calculated in tonnes of carbon dioxide equivalent (tCO_2e).

The average carbon footprint for people varies from country to country.

In industrial nations, the average carbon footprint is about 11 metric tonnes, whereas the average worldwide carbon footprint is about 4.5 metric tonnes. However, the worldwide target to combat climate change is only 2 metric tonnes!

Practical ways to reduce carbon dioxide include:

▶ Reduce the use of fossil fuels by:
  ▷ replacing old incandescent light bulbs with efficient LED (light emitting diode) light;
  ▷ turning off lights/IT equipment when not in use;
  ▷ reducing the use of air-conditioning units;
  ▷ installing insulation;

▷ using public transport (e.g., trains);
▷ holding conference calls rather than travelling to meetings, etc.; and
▷ using more energy efficient electrical equipment.
▶ Replace fossil fuels with biofuels/renewable energy, for example, solar power.

Carbon offsetting

Carbon offsetting is taking an action elsewhere (i.e., plant a tree) to balance out the overall effect of the carbon emissions of a primary action taken (drive a car). This should satisfy additionality: that is, the offsetting action would not be taken by the business as usual, i.e., the business does not plant trees as part of its normal activities.

Advantages and disadvantages of carbon offsetting include:

Table 7.3 Carbon offsetting

Benefits	Limitations
▶ Many large national and international companies have signed up – so could be seen as a marketing strategy	▶ Complexity of calculating a company's carbon footprint – this will take time to collate
▶ Is a way of helping third world countries improve standards	▶ Additional costs used to offset the carbon produced. In some circumstances, it would be better to find ways of reducing the carbon levels within the organisation directly e.g., installing wind turbine or solar panels, rather than paying carbon offsets
▶ Ease of arrangements as carbon credits are easily traded	
▶ Generally more cost effective as market forces determine where reductions in carbon emissions can be made	▶ It can be difficult to prove additionality
	▶ It conflicts with social equity
	▶ The system requires extensive audits

Savings in energy bills and peak load management

Energy efficiency measures include improving the insulation of buildings, travelling in more fuel-efficient vehicles and using more efficient electrical appliances is important and energy conservation; for example, turning off lights and electronics when not in use thereby reducing electricity demand, for example, a television left on standby uses 50% of the electricity it would use during normal operation.

Energy efficiency will:

▶ Reduce emissions of gases such as carbon dioxide, which is viewed as being a cause of global warming.
▶ Reduce emissions of gases such as sulphur dioxide that cause acid rain.
▶ Reduce energy use, which also conserves non-renewable fossil fuels, thereby extending the life of reserves.
▶ Reduce the number of incidents of pollution arising from extraction, refining and distribution of products such as oil.

Global Warming Potential (GWP) is defined as the cumulative radiative forcing – both direct and indirect effects – integrated over a period of time from the emission of a unit mass of gas relative to some reference gas. The Intergovernmental Panel on Climate Change (IPCC) as its reference gas chose carbon dioxide (CO_2) and its GWP is set equal to one (1).

Substance	GWP
Carbon dioxide CO_2	1
Methane CH_4	21
Nitrous oxide N_2O	310

Peak load management is defined as an '*economic reduction of electric energy demand during a utility's peak generation period*'. Peak load management strategies may be undertaken either independently of or in collaboration with a company's utility or Independent System Operator (ISO)/Regional Transmission Organisation (RTO).

Various methods can be adapted to make savings in energy bills. Savings can be derived from:

▶ Eliminating heaters and air conditioning units operating simultaneously in the same space.

▶ Eliminating/reducing the use of portable heaters, installing a simple time switch so they turn themselves off after a designated period, for instance, 30 minutes.

▶ Setting heating systems to be in operation only when the workplace is occupied.

▶ Setting thermostats to 19–20°C for heating.

▶ Using heating systems that are linked to zones or using thermostatic radiator valves – allowing for adjustments to be made in the temperature in different parts of the workplace.

▶ Reducing the thermostats rather than opening windows to reduce the temperature.

▶ Fitting draught-strips around windows and doors.

▶ Installing local instantaneous water heaters where small quantities of hot water are required.

▶ Insulating all hot water tanks, boilers, valves and pipework, etc.

Revision exercise

Explain why energy efficiency is important. (8)

Exam tip

Identify ways your organisation could improve their energy efficiency

---

7.4 Outline the control measures available to enable energy efficiency

---

Key revision points

▶ Insulation, choice of equipment, maintenance and control systems in minimising energy use; supervision

▶ Building design

▶ Fuel choice for transport, energy efficient vehicles, optimisation of vehicle use, car sharing and the use of other options such as teleconferencing

Insulation, choice of equipment, maintenance and control systems in minimising energy use; supervision

Measures available to enable energy efficiency are wide ranging. These will include ensuring equipment is selected for its environmental benefits, including:

▶ Life Cycle Assessment (Analysis) *
▶ Maintenance of the equipment – how often, and by whom?
▶ Control systems – do they operate at energy efficiency?

(* see figure 3.1)

Building design

Energy efficiency measures can be made in buildings. This can be as simple as turning equipment off, turning lights out and closing doors.

Light switches that are motion sensitive can be used (removing the need for someone to remember to switch them off). By replacing inefficient equipment, running costs will be reduced – for example, LED light bulbs save up to 90 per cent more energy than traditional incandescent bulbs.

General maintenance and basic improvements will include insulation of pipes, etc., which in turn will improve energy efficiency.

Additional options include using energy generated from non–fossil fuels (see above);

Control measures available to enable energy efficiency for buildings include:

▶ Initial design and choice of equipment either for fuel type or plant efficiency.
▶ Replacement of inefficient equipment is a good starting point.

▶ Maintenance is essential to keep performance tuned.
▶ Insulation of pipes and spaces.
▶ Avoiding heating empty accommodations.
▶ Heat recovery from hot process water may be beneficial if large volumes are involved.
▶ Light switches that are motion sensitive.
▶ Low energy light bulbs.
▶ Switching equipment off completely when not in use.
▶ Solar heating or power generation.
▶ Staff training should reinforce the process.
▶ External companies offer advice services and will take over energy management on larger sites with their remuneration based on savings achieved.

Fuel choice for transport, energy efficient vehicles, optimisation of vehicle use, car sharing and the use of other options such as teleconferencing

Energy efficiency measures can be made with regard to transport, including: providing driver training; practical measures including using speed limiters, cruise control, GPS systems (for route optimisation); kinetic recovery systems; and tracking devices. In addition maintenance of vehicles and such matters as tyre pressures, etc. will also improve driver safety. The choice of the vehicle and its design features should also be considered. This will include looking at the fuel efficiency of the vehicle. Fuel choice may also be a factor to consider – for example, the use of hybrid cars and electrical cars. Efficiency measures may also include the reduction of travelling – by using conferencing calls/videos, etc.

Alternatives to traveling by road, such as rail for goods or public transport and car sharing for people, may also be considered.

Control measures available to enable energy efficiency for transport include:

▶ fuel selection;
▶ choice of vehicle;
▶ vehicle design features;

- ▶ route optimisation;
- ▶ optimisation of loads;
- ▶ sharing transport i.e. having goods transported for various suppliers/customers;
- ▶ driver training and supervision;
- ▶ use of speed limiters and tracking devices;
- ▶ maintenance of vehicles and such matters as tyre pressures;
- ▶ alternatives to road such as rail for goods or public transport and car sharing for people may also be more efficient; and
- ▶ staff travelling can also be reduced by using telecommunication software.

Revision exercise

Most hospitals use mainly gas and electricity to provide their heating and power. It is essential that patient areas be maintained at reasonable temperatures.

Outline the practical steps the hospital could take to reduce its energy use. (8)

Exam tip

Review your organisation's current driving policy – identify the possible savings that could be made.

EC 1.8

Control of environmental noise

Learning outcomes

8.1 Describe the potential sources of environmental noise and their consequences ☐

8.2 Outline the methods available for the control of environmental noise ☐

8.1 Describe the potential sources of environmental noise and their consequences

Key revision points

▶ The characteristics of noise which lead to it being a nuisance
▶ Sources of industrial environmental noise
▶ Other sources of noise

The characteristics of noise which lead to it being a nuisance

Characteristics of noise include:

▶ Low frequency (can carry long distances).
▶ Speech such as a tannoy i.e. public address systems.
▶ Intermittent such as sirens and explosives.

Noise is generally considered to be any sound that is loud, unpleasant, undesired or unwanted: a nuisance!

Table 8.1 Noise nuisance

Nuisance (private): relating to noise	Nuisance (public): relating to noise
▶ Private nuisance is an unreasonable interference with a person's use or enjoyment of land, or some right over, or in connection, with it	▶ Public nuisance is a crime as well as a tort. It is similar to private nuisance, except that it is well established that there is no need to have an interest in land affected and prescription is not a defence
▶ To be liable under the tort, it should be foreseeable that actions would be likely to give rise to a nuisance	▶ Persons affected are the public, or a section of it, which suffer damage at large
▶ Typical activities actionable under private nuisance include: interference with enjoyment of property	▶ Typical examples of public nuisance would be: blasting noise and flying rocks from a quarry

Noise can be both a private and a public nuisance (affecting an individual or group of individuals/homes/businesses, etc.). This may result in civil or criminal action, depending upon local/national legislation.

Unwanted sound could be caused by factors that determine whether noise from the construction site is likely to constitute a nuisance including:

▶ frequency (low frequencies can carry long distances);
▶ loudness;
▶ within audible range (20Hz–20kHz);
▶ if the noise is speech, such as a tannoy;
▶ time of day and duration;
▶ the character of the neighbourhood;
▶ proximity to habitation;
▶ background noise levels; and
▶ whether there is intermittent noise such as sirens and explosives or impact noise involved.

The effects of noise can be wide ranging. Human health effects include damage to hearing, loss of sleep, and disruption to wildlife. At certain frequencies and intensity, structural damage to buildings is possible.

Rating Industrial Noise Affecting Mixed Residential And Industrial Areas: BS 4142:2014 provides guidance on:

▶ methods for determining noise levels outside of a building;
▶ assessing noise complaints;
▶ equivalent continuous A-weighted sound pressure level, L AeqT;
▶ ambient noise levels;
▶ residual noise levels; and
▶ background noise levels.

A general guide is that an increase of more than 10 dB(A) would be likely to give rise to complaints. BS 4142:2014 is not suitable for assessing noise levels inside of buildings.

Sources of industrial environmental noise

Examples of sources of industrial noise include noise from commercial activities, e.g., machinery, extraction systems, compressor systems, public address systems. In addition workplace noise includes work transport, agricultural/farming, construction, quarrying/mining and workshops/ factories.

Sources of industrial environmental noise include:

- ▶ noise from commercial activities, e.g., machinery, extraction systems, compressor systems, public address systems;
- ▶ transport noise;
- ▶ agricultural noise, e.g., bird-scarers;
- ▶ construction noise;
- ▶ quarrying; and
- ▶ mining.

Examples of construction noise include:

- ▶ site transport/vehicle access;
- ▶ piling;
- ▶ compaction (rollers, vibration, etc.);
- ▶ mechanical plant (excavators, mixers, etc.);
- ▶ deliveries;
- ▶ tools and equipment (drills, saws, riveting, etc.);
- ▶ personnel noise (shouting, etc.);
- ▶ concrete pumping;
- ▶ impact noise;
- ▶ vehicle reversing alarms; and
- ▶ tannoy, radios, etc.

Other sources of noise

Sources of other environmental noise include:

- ▶ noise from pubs and clubs;
- ▶ neighbour noise, e.g., loud music;
- ▶ intruder and vehicle alarms;
- ▶ wind farms; and

▶ local exhaust ventilation systems.

Examples of night-time noise disturbance:

▶ works traffic and in particular employees' cars, forklift trucks, delivery vehicles;
▶ process machinery and equipment;
▶ service equipment and in particular compressors and fans;
▶ general factory noise escaping through open doorways, vents, etc.; and
▶ other intermittent noise sources such as tannoys, radios, employees shouting instructions, etc.

Revision exercise

A large dust extraction fan and collector unit is to be installed against an outside wall of a factory building.

(a) **Identify TWO** possible sources of noise from this equipment. (2)
(b) **Outline** the issues to be considered in order that the equipment does not cause a noise nuisance when in operation. (6)

Exam tip

Identify all types of noise that emanate from your place of work.

8.2 Outline the methods available for the control of environmental noise

---

Key revision points

▶ Basic noise control techniques

BS 4142 sets out the methods for carrying out an assessment of noise arising from an industrial source. Steps that would need to be undertaken include:

▶ Measuring the existing background noise levels. (The 'background' noise level at the assessment location is also measured at a time when the specific noise source is not operating.)
▶ Measuring the noise levels.
▶ Determining the changes in noise emission.
▶ Assessing the actual increase in the levels advised in the standard.

Adjustments may also be required to take account of particular factors such as tonal characteristics and impact noise (if the 'background' noise level at the assessment location is also measured at a time when the specific noise source is not operating). Once recordings have been taken, it is necessary to evaluate the findings, taking into account any other issues such as the prevalence of existing noise nuisance problems and the sensitivity of local residents.

Basic noise control techniques

Basic noise control measures are outlined in Table 8.2.

Table 8.2 Noise control

Basic/engineering noise control measure	
Isolation	Isolation of sound can be achieved by keeping doors and windows closed
Absorption	Absorption of sound can be achieved by installing acoustic linings in the walls of the building
Insulation	Insulation of sound can be achieved by building an enclosure around the source of noise
Damping	Vibrating surfaces should be minimised by reducing the size of panels or fitting material to the panels which reduce the flexibility and consequently the ability of the panel to move

Basic/engineering noise control measure

Silencing	Silencing air emissions can be accomplished by reducing the release of turbulent air into the workplace or by using silencing methods including baffles on exhaust outlets
Environmental noise barriers	Screening the source of noise can help to prevent noise spread. This can be accomplished through

▶ High walls or fences

▶ Purpose-built embankment or bunds

▶ Placing buildings in the vicinity of the noise source

Management controls include:

▶ maintenance of work equipment;

▶ purchasing new equipment with reduced noise production;

▶ resurfacing roads and/or yards;

▶ preventing use of radios;

▶ controlling vehicle noise by managing their use;

▶ controlling hours of working and use of public address systems;

▶ planting trees or building a fence on the boundary;

▶ fitting double and/or triple glazing to the resident's properties if necessary;

▶ increasing the distance (sound decreases by 6 dB for every doubling of the distance away from the source, in open air);

▶ relocating equipment and/or premises; and

▶ referring to BS5228–1 2009+A1:2014.

Maintenance regimes will not only increase the life of the equipment but may also reduce the noise levels.

Revision exercise

(a) **Identify** the main sources of noise from an industrial estate. (4)

(b) **Outline** how noise from the industrial estate could be prevented from interfering with neighbouring residents. (4)

Exam tip

Outline examples of ways to reduce the noise that emanates from your place of work.

EC 1.9

Planning for and dealing with environmental emergencies

Learning outcomes

9.1 Explain why emergency preparedness and response is essential to protect the environment ☐

9.2 Describe the measures that need to be in place when planning for emergencies ☐

9.1 Explain why emergency preparedness and response is essential to protect the environment

Key revision points

▶ General responsibility or duty not to pollute
▶ Part of Environmental Management System
▶ Need for prompt action to protect people and the environment
▶ Risks of prosecution and other costs
▶ Reputational issues

Pollution from a workplace/chemical site can be wide ranging:

▶ affecting human and animal health and damaging plants;
▶ damage to the environment (acid rain);
▶ general nuisance (from odour, noise, etc.);
▶ substances released by 'fallout' causing pollution, land contamination and water pollution; and
▶ pollution may cause damage to property.

Incidents that cause pollution can be very minor, ranging from an oil spill from a work vehicle onto a grass verge to a major chemical release, such as at Seveso, Italy.

Sites that hold large quantities of harmful substances in Europe are required to comply with the current Seveso Directive.

In 1976, in Seveso, Italy, at a chemical plant manufacturing pesticides and herbicides, a dense vapour cloud containing tetrachlorodibenzoparadioxin (TCDD), commonly known as dioxin (a poisonous and carcinogenic by-product of an uncontrolled exothermic reaction), was released from a reactor. Although no immediate fatalities were reported, the substance lethal to humans was widely dispersed. Approximately 2,000 people were treated for dioxin poisoning. This led to the formation of the European 'Seveso Directives', aimed to try to prevent similar incidents. The first Seveso Directive was adopted in 1982.

Seveso II evolved after recommendations resulting from the Bhopal disaster (1987) and Rhine pollution incident (1988). Seveso II included a revision and extension of the scope; the introduction of new requirements relating to safety management systems; emergency planning and land-use planning; and a reinforcement of the provisions on inspections to be carried out by the Member States. This was adopted in 1996.

Seveso III (2012) is a further adaptation of the provisions on major accidents. It includes:

▶ Technical updates to take account of changes in EU chemicals classification.

▶ Better access to information about risks resulting from activities of nearby companies and about how to behave in the event of an accident.

▶ More effective rules on participation, by the public concerned, in land-use planning projects related to Seveso plants.

▶ Access to justice for citizens who have not been granted appropriate access to information or participation.

▶ Stricter standards for inspections of establishments to ensure more effective enforcement of safety rules.

The Member States had to transpose and implement the Seveso Directive by June 2015. To implement this Directive in the UK, the Control of Major Accident Hazards (COMAH) Regulations 1999 (as amended) have been revoked and replaced by the COMAH Regulations 2015.

General responsibility or duty not to pollute

Regardless of the size of the workplace/site, a company has legal, moral and economic reasons not to cause pollution. Legal reasons include having a general requirement to comply with the law to avoid sanctions (in a criminal case) or compensation (in a civil case). Member States of Europe will have to comply with the current Seveso Directive, which in the UK is Control of Major Accident Hazards (COMAH) Regulations 2015. The moral reasons not to pollute have developed from a general duty of care and to meet society's expectations. Economic reasons are straightforward – if pollution was to occur, high levels of fines and clean-up costs would be a likely outcome, therefore management would be wise to reduce the risk of such an outcome by reducing the risk of causing pollution.

The reasons not to pollute include:

▶ Moral reasons
  ▷ duty of care;
  ▷ meeting society's expectations; and
  ▷ progress to sustainable business.
▶ Legal reasons
  ▷ general requirement to comply with the law; and
  ▷ avoidance of sanctions or compensation.
▶ Financial reasons
  ▷ cost savings through reduced bills for materials energy and waste disposal;
  ▷ lower tax charges;
  ▷ supply chain pressures; and
  ▷ relations with investors, insurance companies, etc.

For example, a workplace fire could have a detrimental impact on the environment, with the release of:

▶ acid gases;
▶ ecotoxic substances to the atmosphere and their potential impact on plant and animal communities;
▶ global warming gases;
▶ harmful substances to the atmosphere and their potential human health effects;
▶ ozone depleting substances; and
▶ substances that can create nuisance effects, including smoke and odour.

Significant impact may include damage to:

▶ the land/groundwater;
▶ water resources through escape of polluting substances to surface water or groundwater, i.e., from fire fighting water runoff;

▶ polluting substances entering foul sewers and on sewage treatment systems;

▶ adjoining properties through damage caused by explosion or radiant heat; and

▶ generation of large amounts of contaminated waste that would require disposal after the fire.

Part of Environmental Management System

Many organisations have implemented an Environmental Management System (14001). Within the current management system under Implementation and Operation (4.4), there is a requirement to plan for an emergency (4.4.7 Emergency Preparedness and Response).

Need for prompt action to protect people and the environment

It is imperative that the health and safety of people (those working or visiting the place of work and those who may be affected by any environmental emergency) are prioritised over damage to the environment. Likewise, the organisation must also protect the environment.

Action to be taken in the event of a fire:

▶ Contact emergency services.
▶ Monitor firewater runoff – to local rivers, etc.
▶ Use drain covers, booms, absorbent materials to reduce the impact of firewater runoff.
▶ Contact relevant agencies/government bodies (e.g., Environmental Agency).
▶ Isolate known surface drains or divert to holding ponds.

Action to be taken in the event of an oil spill:

▶ Take immediate action to stop the spillage of oil:
  ▷ Identify the source.
  ▷ Stem the flow.

> ▷ Prevent the spread.
> ▷ Report.

▶ Gather details of oil spillage.

▶ Monitor any possible discharge points, and identify if oil has been spilt.

▶ Contact relevant agencies/government bodies
(e.g., Environmental Agency).

Environmental emergency response plan:

▶ Plans are required as part of an EMS under ISO14001.

▶ Legal requirements – for example, in the UK they must comply
with Environmental Permitting (England and Wales) Regulations
2010 SI 675 (the old standard was the integrated pollution
prevention and control [IPPC] (EC 2008/1/EC).

▶ Demonstrate examples of good moral/ethical standards.

> ▷ It is the 'right thing' to do, i.e., look after workforce,
> neighbours and environment?

▶ A plan helps meet the need for prompt action to protect people
and the environment in the event of an incident.

▶ It should reduce the risk of prosecution.

▶ It should reduce clean-up costs if appropriate.

▶ It should help prevent reputational damage and loss of business.

▶ It should include contingencies to prevent water pollution due to
fire or another catastrophe.

Example of actions in the event of an oil spill:

▶ Information gathering such as details of the complainant and the
location and time of observation.

▶ Inspecting the watercourse, outfalls and discharge points, oil
storage facilities and interceptors and ascertaining if there have
been deliveries or movements of oils.

▶ Environmental Agency and Water Company should be advised.

▶ Key internal contacts including emergency response team should
be alerted.

▶ Contact known downstream activities/organisations.

▶ Initiation of emergency plan including:

> ▷ access, locking off the plant or sealing drains; and
> ▷ deploying spill and emergency kits.

▶ Taking samples of discharges and watercourse.

Typical emergency response arrangements that should be in place to minimise environmental harm:

▶ Fire and firewater runoff risk assessment.
▶ Development and maintenance of emergency response plans.
▶ Arrangements for fire detection, response and firefighting.
▶ Command and control arrangements.
▶ Provision of emergency response equipment, including firefighting, drain covers, booms, absorbents and allowance for firewater containment.
▶ Off-site emergency plans.
▶ Arrangements for notifying relevant organisations.
▶ Arrangements for practising emergency plan through desktop rehearsals or simulated incidents.
▶ Need for regular training of staff and contractors in emergency response procedures.
▶ Stage reporting and recovery systems.

Risks of prosecution and other costs

In the event of a successful prosecution after an environmental emergency, the tangible costs would include:

▶ legal – for solicitors and expert witnesses;
▶ fines – imposed by the courts;
▶ clean up – to 'put right', if possible, put back the environment as it was before;
▶ compensation – payable to local residents and others affected by the environmental emergency; and
▶ time costs of all people involved in the clean-up, environmental redemption, investigation.

Reputational issues

Whilst it will be possible to make an accurate calculation of the tangible costs, i.e., possible prosecution, clean-up and other costs, the intangible costs to an organisation's reputation and goodwill will be far higher as a result of an environmental incident and will be almost impossible to calculate.

169

Revision exercise

Outline the key actions that need to be taken to protect the environment in the event of a serious fire breaking out on industrial premises. (8)

Exam tip

Review the findings of a major environmental incident/ emergency (for example, Buncefield www.hse.gov.uk/comah/ buncefield/buncefield-report.pdf).

9.2 Describe the measures that need to be in place when planning for emergencies

Key revision points

- ▶ Emergency response plan (to include foreseeable internal and external causes)
- ▶ Emergency control centre
- ▶ Training and practices
- ▶ Recognising risk situations and action to take
- ▶ Materials to deal with spills
- ▶ Access to site plans
- ▶ Inventory of materials
- ▶ Environmental hazards associated with fire
- ▶ Liaison with regulatory bodies and emergency services
- ▶ Protecting and liaising with the local residents, including indigenous peoples
- ▶ Handling the press and other media

Emergency response plan (to include foreseeable internal and external causes)

The objectives of an emergency plan are to contain and control major incidents and by implementing the necessary measures to protect persons, property and the environment from their effects. Additionally, the emergency plan has a part to play in communicating necessary information to the public, emergency services and area authorities and in providing for the clean-up and restoration of the environment following a major safety/environmental incident. For sites that fall under the Seveso Directive, emergency plans are required to be reviewed, tested, revised and updated at least every three years.

Information that should be included within an emergency plan is as follows:

▶ The name or position of the person in the organisation with the responsibility for liaising with regulatory bodies and emergency services.

▶ The names or positions of the persons in the organisation with the authority to set the emergency procedures in motion.

▶ The name or positions of the persons in charge of coordinating the on-site mitigatory action: the action is taken to control foreseeable conditions and events and for limiting their consequences.

▶ The safety equipment and resources available.

▶ The methods for giving warnings to persons and the actions expected of them on receipt of the warning.

▶ The arrangements for training staff in the duties they will be expected to perform.

▶ The arrangements for limiting risks to persons on site.

▶ The arrangements for giving early warning of an incident to the local authority responsible for setting the off-site emergency plan in motion and the type of information that should be contained in the initial warning.

▶ Arrangements for the provision of more detailed information as it becomes available.

▶ Coordinating staff training with the emergency services.
▶ Development of arrangements for providing assistance with off-site emergency action.
▶ Specific details will also be included: details of the site operations; location of nearest neighbours; location of Sites of Specific Scientific Interest; drainage plans and location of emergency equipment (including Personal Protective Equipment, first aid and firefighting equipment).
▶ How to alert neighbours.
▶ Recovery plans.
▶ Finally details about how to deal with the media and local businesses and residents.

Emergency control centre

For sites that fall under the Seveso Directive, Emergency Control Centres (ECC) are required, from which the emergency response operations are directed and coordinated. The ECC will be set up with internal and external communications equipment. The ECC will also have details of the site, including plans and maps, clearly showing the location of hazardous materials; the amount of stocked hazardous material on site; location of transport facilities; location of facilities to assist emergency services; assembly points and casualty treatment areas, etc.

Training and practices

It is necessary to ensure that all procedures and systems are functioning. This will include the need to undertake regular testing and practices of the emergency systems in place. Arrangements for specific training including ongoing and refresher training for key personnel involved in the emergency process is required. All emergency plans prepared in accordance with Seveso Directive, where applicable, should be tested both on-site and off-site at least every three years. The tests conducted should be based on an emergency scenario that has previously been identified in the assessments of the likelihood of the occurrence of such emergencies on the installation. A range of activities will be used as part of the training process; this will involve: drills, table top exercises, walk through exercises (the response is walked through, including visiting specific aspects of the plan) and live exercises, fully testing some or all aspects of responses. In addition, there will be seminar exercises and controlled post exercises, testing the communication arrangements during an emergency and facilitated discussions about the organisational roles and responses in the various circumstances of an emergency.

Recognising risk situations and action to take

There is a wide range of emergencies that an organisation may need to deal with. As mentioned above this could be anything from a small oil spill to a major release of toxic gases or fumes. It may be necessary to plan for the following likely emergencies. Causes of environmental incidents on your site may include failure of equipment; e.g., containment failure of tanks, pipes (from wrong connections of sewers and pipes, etc.); plant failure; or equipment failure. Procedural failures would include: overfilling containment vessels; spillage during the delivery and use of materials; accidental discharge of partially treated or raw effluent; or failure to contain firefighting water after a fire/explosion. Natural disasters such as flooding and human actions such as vandalism could also cause damage to the environment.

Materials to deal with spills

The provision of a suitable spill kit in case of emergencies should be made available and must be appropriate for the substances they are to clean up. Training must be given. There is a range of different types of spill kits to cope with a number of different types of substances, including oils, acids and alkalis. A typical spill kit will contain gloves, absorbent materials, shovels, brooms, plastic bags, etc.

Disposal of substances should be undertaken correctly – following manufacturer's guidance.

Other equipment needed will include drainage covers and seals, absorbent or inflatable booms, portable tanks and sand that may be used, in some incidents, as an absorbent medium.

Access to site plans

Various organisations may need to have access to site plans in the event of an emergency, such as emergency services, environmental regulators and the health and safety regulators. All workers, visitors and those who may be affected may also need access to the site plans.

Features that should be included in site plans are:

▶ access and egress details;
▶ schematic representation of the site drainage arrangements;
▶ layout of buildings;
▶ access routes and meeting points for emergency services;
▶ the location of process areas and any on-site treatment facilities for trade effluent or domestic sewage;
▶ areas or facilities used to store raw materials, products and wastes;
▶ bunded areas, with details of products stored and estimated retention capacity;
▶ location of emergency equipment including hydrants, 'fireboxes' and pollution prevention equipment and materials;
▶ details of watercourses, springs, boreholes or wells located within or near the site;
▶ areas of porous or unmade ground;
▶ site drainage – foul, surface and trade effluent drainage systems; and
▶ any known sites of Special Scientific Interest.

Information that should be contained within an environmental emergency response plan include details of:

▶ company location;
▶ site plan;

174

- access and key staff with contact details;
- site operations;
- location of nearest neighbours;
- drainage plans;
- control centre;
- location of emergency equipment;
- external contacts such as fire service, Environmental Agency, sewerage undertaker;
- dealing with the media and local businesses and residents;
- clean-up processes;
- waste disposal;
- information about authorship;
- review dates;
- staff training;
- exercise/training records; and
- details of Special Scientific Interest.

Steps to ensure the effectiveness of an environmental emergency response plan:

- Training for new staff and refresher training for existing staff.
- Ensuring that everyone is aware of their responsibilities.
- Practices and drills – involving neighbours and primary responders as appropriate.
- Regular liaison with those who may be involved.
- Any real incidents or alarms need to be evaluated and any shortcomings in plan rectified.
- Regular reviews of plans and updating plans if circumstances change.
- Maintenance and testing of systems and alarms.
- Emergency signage, instructions to visitors, emergency equipment and personal protective equipment (PPE) all need to be regularly checked.
- Copies of the plan should be held off-site and in the emergency control centre.
- Emergency control centre should be clearly identified and equipped.
- The plan should be up-to-date with contact details of both internal and external parties.
- An external audit of the plan should build confidence.

175

Emergency training may include:

▶ Awareness of potential harm from substances on site.
▶ Awareness of sensitivity of environment surrounding the facility.
▶ Use and limitations of PPE.
▶ Reporting procedures for contacting relevant agency and/or water undertaker if there is a risk of pollution.
▶ Clean-up, safe handling and legal disposal of contaminated materials and wastes.
▶ Arrangements for use of specialists, contractors and services.
▶ Appropriate decontamination or legal disposal of contaminated waste and/or land.
▶ Re-stocking procedure for emergency equipment.

Inventory of materials

The site plan should provide details of all stores, bulk storage vessels, drums or containers used for storing oils, chemicals or other potentially polluting materials. A detailed inventory should be kept – this will need to be updated regularly.

For example, in England, this is covered within the Oil Storage Regulations Control of Pollution (Oil Storage) Regulations (England) 2001.

Environmental hazards associated with fire

Environmental impacts that are likely to arise from a major fire will include the release of:

▶ Harmful substances to the atmosphere and their potential human health effects.
▶ Ecotoxic substances to the atmosphere and their potential impact on animals and plants.
▶ Fallout of polluting substances onto land, water, crops or buildings.
▶ Ozone-depleting substances and global warming gases.

176

▶ Substances which can create nuisance effects, including smoke and odour.

They may also have a detrimental impact upon water resources through escape of polluting substances to surface water or groundwater. There is a potential to generate large amounts of contaminated waste that would require disposal after the fire. The polluting substances that enter foul sewers will have a detrimental impact on sewage treatment systems. The impact of a major fire could cause damage to adjoining properties by explosion/radiant heat.

Liaison with regulatory bodies and emergency services

In the event of an emergency, it will be necessary for management to liaise with:

▶ emergency services;
▶ the environmental and/or health and safety regulator;
▶ the local and/or national authority; and
▶ the local water company.

The emergency response plan will include details of the name or position of the person in the organisation with the responsibility for liaising with regulatory bodies and emergency services. It is important that only the authorised persons liaise with the regulatory bodies and emergency services, to prevent miscommunication of information and to decrease the risk of criminal legal action.

Protecting and liaising with the local residents, including indigenous peoples

A designated person should be made responsible for communicating with local residents, including indigenous peoples (where applicable). It is important that only the authorised persons liaise with the local residents, including indigenous peoples (where applicable), to prevent miscommunication of information and to decrease the risk of civil legal action.

177

Handling the press and other media

Likewise, a designated person should be made responsible for communicating with the media. It is important that only the authorised persons liaise with the media, to prevent miscommunication of information and to decrease the risk of civil legal action.

Revision exercise

(a) **Outline** why an environmental emergency response plan is advisable for large manufacturing sites. (10)

(b) **Outline** the information that should be contained in an emergency response plan and the information that should be available to support the plan. (10)

Exam tip

Outline emergency response arrangements at your place of work.

Answering NEBOSH type exam questions

The following question shows a practical example of each of the command words that are used by NEBOSH (answering a non-environmental management question).

Question:

(a) Identify types of fruit. (2)
(b) Give an example of a fruit that would not be used in a fruit salad. (1)
(c) Outline steps to make a fruit salad. (2)
(d) Describe a banana. (2)
(e) Explain the benefits of eating five fruits and/or vegetables per day. (1)

The following shows model answers to the above EIGHT mark questions, following notes on how marks are awarded:

(a) Types of fruit.
  ▶ apple;
  ▶ pear;
  ▶ kiwi;
  ▶ orange.

(b)

▶ Tomato.

(c)

▶ Wash fruit that does not need to be peeled.
▶ Cut the fruit into small pieces.
▶ Place the pieces into a large bowl.

(d)

▶ The bottom end is narrowed; the top has a thick stem, attaching the fruit to the stalk.
▶ Ranges from about 10 to 20 cm in length and is curved/crescent in shape.
▶ The colour changes from green (unripe), to yellow (ripe), to black (over ripe).
▶ Has a thick skin that needs to be removed before eating.

(e)

▶ Provides a range of vitamins and minerals.
▶ Provides fibre within a diet.

Notes

1 You do not need to write out the question before answering it.
2 You only need to have the number of answers per marks given in the question. For example, (a) identify types of fruit offers two marks, so apple and pear were all that was needed.
3 If you put extra answers that are similar; for example in (a) you wrote 'Braeburn, Red Delicious and Gala' this is unlikely to have gained the two marks even though three answers were given that were all types of apples.
4 If the question is an Outline, Describe or Explain, then a short sentence will be required for the answer. If you had only written 'Wash and cut fruit', whilst this is correct it would not have gained the two marks.
5 If you get an answer wrong and then one that is correct, you will not lose mark(s); for example, for part (b) if you had written 'grapes, tomato' then you would still have gained the mark needed, for tomato.

✎ **Notes**

✎ **Notes**

Index

Index

Printed and bound by CPI Group (UK) Ltd, Croydon, CR0 4YY

01/11/2024

01782634-0001